Studies in Smart Technologies

Advanced communication and computing technologies to develop intelligent and adaptive systems are usually referred to as smart technologies. Many of the sustainable development goals (e.g., SDG 4, SDG 6, SDG 7, SDG 8, SDG 9, SDG 11) cannot be achieved without developing smart technologies. The aim of the proposed book series, "Studies in Smart Technologies," is to provide a detailed and latest overview of the latest research findings in the field of smart technologies and allied areas. The series covers studies related to smart cities, IoT, smart sensors, AI, soft computing, and other related areas. The content published in the series will be an important reference to the policymakers, industries, researchers, practitioners, and students in computer science, electrical engineering, information technology, and aerospace engineering.

The series will publish monographs, edited volumes, textbooks and proceedings of important conferences, symposia, and meetings in the field of smart technologies and their applications.

Tanvir Habib Sardar

The Spectrum of Computer Science: Emerging Technologies and Trends

Springer

Tanvir Habib Sardar
Department of Computer Science
and Engineering
School of Engineering
Dayananda Sagar University
Bengaluru, Karnataka, India

ISSN 3004-9881 ISSN 3004-989X (electronic)
Studies in Smart Technologies
ISBN 978-3-032-19631-6 ISBN 978-3-032-19632-3 (eBook)
https://doi.org/10.1007/978-3-032-19632-3

This Springer imprint is published by the registered company Springer Nature Switzerland AG
The registered company address is: Gewerbestrasse 11, 6330 Cham, Switzerland

Foreword

In an era where computing technologies evolve faster than ever, understanding their interconnections is essential for both learners and professionals. *The Spectrum of Computer Science: Emerging Technologies and Trends* captures this evolution with clarity and coherence. Dr. Tanvir Habib Sardar presents a well-structured exploration of the major disciplines defining modern computer science—from artificial intelligence and data science to quantum computing, blockchain, and the metaverse.

What sets this book apart is its balance between foundational principles and forward-looking insights. Each chapter invites readers to reflect not only on the technical mechanisms but also on their ethical, societal, and human implications. The lucid explanations and structured organisation make complex topics approachable without diluting their depth or rigour.

For students, it provides a roadmap through the rapidly expanding landscape of computing concepts. For educators and researchers, it serves as a cohesive synthesis of contemporary developments that stimulate further inquiry and classroom discussion. In today's world, where digital transformation is redefining how we work, learn, and collaborate, such a comprehensive yet accessible perspective is both timely and valuable.

Dr. Sardar's thoughtful presentation reflects his years of experience as a teacher and researcher, bridging theoretical understanding with practical insight. This book will undoubtedly serve as a guiding resource for those seeking to comprehend not just what computing is today, but what it is becoming in the decades ahead.

Dr. Saurav Mallik
Research Scientist III
Department of Pharmacology and Toxicology
R. Ken Coit College of Pharmacy
The University of Arizona
Tucson, USA
smallik@pharmacy.arizona.edu

Preface

This book is written with the reader in mind—to serve as a guided journey through the technologies, techniques, and algorithms shaping today's computer world. Each chapter stands independently, yet together they build a seamless flow from fundamental principles to advanced ideas that define the digital transformation of society.

The book is divided into 14 chapters, covering emerging areas such as generative AI, quantum computing, data analytics, machine learning, blockchain, IoT, and the metaverse. Each chapter introduces core concepts; technological foundations; and their implications for industries, research, and humanity as a whole.

My objective in writing this book is twofold: to make complex concepts accessible to students and to provide professionals with a consolidated understanding of emerging trends. The language has been carefully designed to strike a balance between academic rigour and simplicity, ensuring that readers from diverse backgrounds can engage meaningfully with the material.

I hope this book serves not only as a reference but also as an inspiration—encouraging readers to explore, innovate, and contribute to the ever-expanding frontier of computer science.

Dr. Tanvir Habib Sardar
Associate Professor
Department of Computer Science and Engineering
School of Engineering
Dayananda Sagar University
Bengaluru, Karnataka, India

Acknowledgements I would like to express my sincere gratitude to all who have supported and inspired me throughout the development of this book.

My heartfelt appreciation goes to my former students at GITAM (Deemed to be University), Bengaluru, whose curiosity, enthusiasm, and constant desire to explore emerging concepts in computer science inspired me to begin this writing journey. Their questions and engagement in class discussions often sparked new ideas and motivated me to present complex topics in a more structured and accessible way.

I also extend my deep thanks to my current institution, Dayananda Sagar University, Bengaluru, for providing a supportive academic environment and the flexibility that allowed me to work seamlessly, often late into the night. The university's encouragement towards academic and research pursuits has been instrumental in the successful completion of this work.

I extend special thanks to my family for their constant encouragement, patience, and understanding throughout the writing process. Their support has been my greatest source of strength.

Finally, I wish to acknowledge the countless researchers, educators, and innovators whose dedication continues to push the boundaries of what computers—and humans—can achieve together.

Competing Interests The author has no competing interests to declare that are relevant to the content of this manuscript.

Contents

About the Author

Dr. Tanvir Habib Sardar is Associate Professor in the Department of Computer Science and Engineering, School of Engineering, at Dayananda Sagar University, Bengaluru, India. With over 17 years of academic and research experience, he specializes in big data, machine learning, distributed computing, artificial intelligence, and emerging computing paradigms. He has published 81 Scopus-indexed research articles, authored and edited over ten books, and holds more than a dozen patents.

Dr. Sardar's current research focuses on scalable machine learning models, including quantum computing, quantum-optimized intelligent systems, neuro-fuzzy computing, and deep learning, alongside distributed frameworks such as Hadoop. His work bridges theoretical foundations with real-world application, fostering innovation, problem-solving, and critical thinking among students and researchers. Beyond his scholarly contributions, he actively engages in curriculum development, accreditation processes, and mentoring initiatives that support future-ready engineering education. e-mail: tanvir-cse@dsu.edu.in

Chapter 1
Introduction to Modern Computer Science

1.1 Understanding the Scope of Computer Science

The field of computer science is a broad field. Computer science can be divided into different subfields, and each of these subfields can address many applications [1]. The foundation of computer science is actually based on mathematics. The study of different kinds of computational processes, the management of information, the automation through digital systems, etc. However, the scope of computer science goes beyond all these fundamental components. For example, in today's world, the traditional data is now changed to big data. So big data is a new domain of computer science where we analyse big data. To get insight from the data and to suggest new solutions for the businesses, governments, and different kinds of agencies. Similarly, we have human–computer interactions, and we have software engineering, artificial intelligence, machine learning, quantum computing, and many of its fields which are coming up in the future and is populating the field of computer science.

One of the basic concepts in computer science is programming [2]. Sometimes it is known as the gateway to computer science. In general, programming is considered synonym of writing code. However, writing code is not the entire thing of programming. The programming is use of proper scientific and mathematical ideas while analysing proper scientific and mathematical ideas while analysing its logical inherent problems to provide scientific solutions and to develop algorithms. These algorithms will design complex systems so that they become scalable, efficient, sustainable, and can provide real-world solutions. The different subfields of computer science and their diversities reflect the interdisciplinary nature of computer science. Over time, it produces the required inventiveness while merging the empirical research aspect with the practical application.

T. H. Sardar, *The Spectrum of Computer Science: Emerging Technologies and Trends*, Studies in Smart Technologies, https://doi.org/10.1007/978-3-032-19632-3_1

1.2 Historical Context and Evolution

From the mechanical computing equipment that made up the initial generations of computers to the contemporary software-powered computing [3] systems, which are adaptive solution providers, there has been an enormous change in the historical background and progression of computer science. The early pioneers of computer science, for example, Ada Lovelace, who actually first conceptualised the idea of algorithms which can be designed for a computing machine. And so we combine Alan Turing, who theorised computations that formed the basis of modern algorithms and artificial intelligence, and so on. They were highly influential in shaping the concept, form, and structure of the modern computer science [4]. Computer science has an extensive history that is created through the efforts and perseverance of thousands of workers and scientists around the world.

The first general-purpose electronic digital computer was ENIAC and was created by Leslie during his residence in Philadelphia. These were hardware-oriented computers. The direction in which computers have been designed has changed over the years as the primary concern has been the software and not the hardware. Though the hardware factor was also considered, in that way the modern computer hardware could not be compared with the previous computers. Today the software-based creation of computer science has been concentrated on offering intelligent remedies through the utilisation of current technologies, techniques, and programming concepts with the help of machine learning, quantum computing, neural networks, deep learning, computer vision, and numerous other concepts. Such developments are attributable to the interdisciplinary character of computer science and reflect its concern with addressing problems in society.

1.3 The Importance of Understanding Computer Science

Computer science is not a notion that has to be applied in the office only in the twenty-first century. It has infiltrated all spheres of human life, in the workplace or in everyday life, in the life of a child or in the life of an old man. Therefore, it is important for all segments of the population be equipped with an elementary awareness of the field of computer science. The more recent notion of artificial intelligence, which is an invaluable component of modern computer science, has an impact on people's lives in a variety of social and professional contexts.

The design and implementation of an IT infrastructure, the administration of a sizable database, the safeguarding of digital resources, or simply its interactions with the digital world, as perceptive citizens all depend on computer science. It improves moral consciousness, creativity, and problem-solving skills—all of which are becoming more and more crucial in a world that is changing quickly.

1.4 The Focus of This Chapter

It is a chapter that is used as a basis to provide insight into the field of computer science. It outlines its core concepts, primary fields, and historical development. Topics such as cybersecurity, software engineering, artificial intelligence, and theoretical computer science will become more clearly understood by readers.

This chapter also introduces key programming paradigms, languages, tools, and fundamental principles of data structures and algorithms. These elements provide readers with the conceptual and technical groundwork necessary for further study and serve as a foundation for more in-depth exploration of the discipline in subsequent chapters.

1.5 Why It Matters

Computer science is the engine that runs the computer world. Everywhere in the world, especially in this highly technological world, computers advance in almost every industry: healthcare, finance, education, and entertainment. Understanding the basics behind computer science gives you the power to be part of this ever-evolving scene. No matter what your goals are, whether becoming a software developer, a data scientist, a cybersecurity expert, or just being a knowledgeable citizen in the twenty-first century, there is plenty of computer science that will provide you with the tools you need to be successful.

Computer science spurs technological advancement in the medical field, finance, education, and entertainment fields. Modern life depends on its principles on the systems of automated diagnosis in medicine to safe money transfer and personalised learning platforms. The knowledge of computer science gives people the power to be active in the digital era, either by taking part as software creators, analysts of information, or individuals who can implement scientific elements of cyberattacks and prevent them either professionally or personally. It opens possibilities to creativity, critical thinking, and impact on society, providing resources to deal with complicated problems and imagine new ones.

This book and its chapters will enable readers to explore the wide landscape of computer science. They will first come to know about the evolution of computer science. The objective is to cultivate curiosity, to inculcate innovation, and to know the foundation of computer science. The readers will come to know about different interdisciplinary roles and applications of computer science, which are shaping the future of human beings. It will introduce the readers to the recent themes, tracks, ideas, and concepts of computer science which will impact our life in the very near future. The world of possibilities will be clear to the reader when they go through the existing and emerging topics of computer science.

1.6 Overview of the Evolution of Computer Science

The history of computer science is relatively very less in time. It started in the middle of the twenty-first century. But arguably, it is one of the most significant technological changes. Based on the foundation of mathematics, electronics, and computational logic, the field of computer science has developed very quickly compared to any other technological field. It is to be mentioned that it is even progressing at a faster speed. This section elaborates on how this interdisciplinary field has progressed throughout its historical trajectory and illuminates how foundational theories are transformed into real-world problem-solving systems. Understanding this evolution is essential for appreciating the scope of computer science today and contextualising its future directions.

1.6.1 The Origins of Computer Science

Computer science, in fact, began in mathematics and logic, concepts whose history goes back centuries before the modern computer was invented. The idea of an algorithm, a sequence of calculations, can be originated in antique civilisation with the help of a working definition. Mathematicians including Euclid who contributed to systematising ways of making geometric proofs, and mathematicians later, including the historical AI-Khwarizmi who brought about algebra itself, established a foundation of algorithmic thinking [5].

1.6.1.1 Mathematical Theories and Algorithms

The notion of an algorithm, which is the finite, ordered series of steps in solving a problem, has its origins in such ancient theorists as Euclid, whose algorithm of computing the greatest common divisor is still fundamental. During the Islamic Golden Age, the Persian mathematician Muhammad ibn Musa al-Khwarizmi formalised some procedures of algebra that would diminish to form the basis of algorithm design; his name forms the etymological basis of the word algorithm. The seventeenth and eighteenth centuries saw the construction of mechanical calculators, including the Pascaline and the stepped reckoner invented by Blaise Pascal and G.W. Leibniz, respectively. Even primitive, these machines represented the principle of mechanised computation and prefigured the subsequent programmable machines [6].

1.6.1.2 Pioneers of Computer Science

During the nineteenth century when the computer pioneers emerged into the mainstream, Charles Babbage was viewed as the father of computing and a British mathematician whose innovativeness and mind projected in the design of the Analytical Engine, which was seen to be the first mechanical general-purpose computer that would perform any calculation or algorithm. It was never constructed, yet the design produced by Babbage also has certain basic elements of modern computer technology, the CPU and memory, just to mention two [7].

The first computer programmer to the world is commonly considered to possess been Ada Lovelace, who worked with Babbage. She authored the earliest algorithm that was to be performed in a computer. Her work imagined the future of the computers not only as number-crunching machines; she foresaw time to come when the computers would be used in the realm of artistic expression, music, and, in fact, much more, and, nowadays, it is already starting to happen in such areas as artificial intelligence, and even creative content development.

1.6.1.3 The Turning Machine

The 1930s concept of the Turing machine encouraged by Alan Turing in the 1930s is the ability of a theoretical construct to model any algorithmic process. This model gave a formal definition of computation and the basis of the study of theory of computation.

The publication of his work covered the major issues of computability and contributed to the design of digital computers, and set the principle of universality that forms the basis of modern programming languages and systems [8].

1.6.2 The Development of Early Computers

The twentieth century brought on the big turn from theory to practice in computation. Machines, which came to define the age of the modern computer. The early computers were driven by the need to automate complex calculations, mainly in such areas as this encompasses science, engineering, and cryptography [9].

1.6.2.1 Mechanical and Electromechanical Computers

At the start of the twentieth century, mechanical and electromechanical computers were developed and were primarily used in special purposes, including codebreaking and scientific calculations. The earliest example of this is the Zuse Z3 which was built in 1941 by the German engineer Konrad Zuse. The computer is considered the pioneer programmable digital computer.

During World War II, British cryptographers developed the Colossus; this was an electromechanical computer that was able to crack the German codes. The success of the Colossus was the first that demonstrated how a computer could accomplish large volumes of automated and high-speed mathematical calculations in a significantly more efficient way than human operators.

1.6.2.2 The ENIAC and UNIVAC

The Electronic Numerical Integrator and Computer, or ENIAC, as a short name, was a general-purpose digital computer capable of carrying out an unimaginable number of computations in previously unimaginable speeds, which was released by the United States in 1945 to serve as a priceless asset to science and military purposes.

As a matter of fact, the initial commercially built computer in the United States was the UNIVAC I or the Universal Automatic Computer. It was a success after the proper foresight on the turnout of the 1952 presidential election in the United States. It was also therefore the beginning point of the computer era in the life of the business people as well as in the minds of people.

1.6.3 The Evolution of Software and Programming

With the growth of the hardware capabilities, the demand to have more efficient and accessible programming paradigms was introduced. The invention of software and programming languages turned computers into specialised machines to be highly adaptable and innovative.

1.6.3.1 From Machine Codes to High-Level Languages

The development of machine codes into high-level programming languages is one of the most crucial changes in the history of computers. The first computers used binary machine code, sequences of 0s and 1s, as instructions to the hardware directly. Although it was powerful, it was tedious, error-prone, and was not available to a lot of users. This was slightly enhanced through the assembly language which encoded instructions in the form of mnemonics but still had to be fully familiar with hardware architecture. In the next stage of development, a new set of programmer-friendly programming languages came into the picture. These programming languages are known as high-level programming languages. These programming languages provide a set of English language constructs known as syntax to the programmers for coding using a predefined structure. This syntax is transformed into computer-specific machine code after "compilation" by the programming language. Compilers basically transform the high-level structured syntax into machine-specific code to close the gap between human reasoning and machine execution. The easily readable syntax improves the programmer's productivity and scalability. As a result, programmers start creating user-friendly software at a greater speed and accuracy with a greater look and feel.

1.6.3.2 The Transitions of High-Level Languages

Early computer programmers would communicate with the machine in low-level machine code, which consisted of binary code (0s and 1s) that were directly translated into hardware instructions. This was a very accurate but also cumbersome and error-prone technique that required a detailed understanding of how the computer worked, providing minimal abstraction and flexibility. As computing needs grew, assembly language emerged as a modest improvement. It introduced mnemonic codes and symbolic labels to represent machine-level instructions, which allowed programmers to create more readable code.

However, the challenge with using assembly languages was that they were dependent on specific hardware configurations. This restricts portability and results in an extremely slow learning curve for programmers. The second paradigm shift occurred in the 1950s and 1960s with the publication of high-level programming languages. High-level programming languages like Fortran and COBOL provide the programmer with the scope to develop scientific and business-specific applications, respectively. The high-level programming language named "BASIC" made learning high-level programming language easier for the students of educational institutions. The high-level programming languages create an abstraction for the programmer such that they don't need to know the hardware specification and machine-specific code. Rather, programmers can now only focus on the syntax and the structure of the programming language. This has reduced the complexity of learning a programming language and increased the productivity of the programmer. This removal of complexity and making the programming at the syntax and structure level made the high-level programming languages very popular for the masses and the programmer alike.

1.6.3.3 The Impact of High-Level Languages

The software development ecosystem has been totally revolutionised by high-level programming languages. Using these high-level programming languages, it has become much easier to create complex and powerful applications than ever before. These languages have removed the complexities of knowing the machine-level instructions and hardware-specific complications. The simplicity of high-level programming languages allowed the developer to develop quality code for scientific research, business analytics, educational technologies, as well as for government organisations and their requirements. The abstraction provided by the high-level programming language allows for applications to be executed on multiple systems without worrying about hardware complexities. It has made it possible for engineers to create software that will operate on various hardware devices across the entire world. A greater degree of versatility has been made feasible by this, and a new branch of software design emerged. These days, logical design, adaptability, and maintenance have become vital in this area of software engineering. The key factor in this revolution of high-level programming languages was the compiler. These compilers can compile human-readable high-level syntax and structure-oriented code into machine-executable code. Therefore, the programmer can write the code once, and after compilation, it can be executed on any computer with the same compiler installation. The programmers could now focus on logic, architecture, and user experience rather than low-level system mechanics. This shift accelerated the rate of innovation and opened up software development to all professionals across disciplines and encouraged computing as an interdisciplinary team effort.

1.6.4 The Rise of Personal Computing

It was with the arrival of the personal computer that democratisation of computing actually began. Technological breakthrough of the microprocessor and strategic advances in the software and user interface design brought the power of computing to homes, schools, and small businesses [10].

1.6.4.1 The Personal Computer Revolution

1980s personal computer revolution was a groundbreaking era in the history of computing, as it broke the exclusive job of computing power, which was left to the specialist laboratories, and brought it into the homes, schools, and small enterprises. This was made possible due to the rapid innovation within the area of microprocessor and led by the visionary companies such as Apple and Microsoft. One of the biggest turning points was Apple's 1977 introduction of Apple II, which was among the first commercially successful personal computers available. The Apple II was simpler and more accessible to use and exploit, especially in learning and business, with its intuitive interface and included programs. By early 1980s, Microsoft had already launched the operating system of the IBM PCs, MS-DOS that rapidly became the standard of personal computing. Its popularity brought standardisation in the development and use of software in various systems. The next release of Microsoft windows brought in the use of graphical user interface whereby the command line inputs were substituted by visual icons and windows. This technology reduced drastically the entry barrier of the non-technical users and made computers a useful tool in their daily chores. PC revolution democratised access to technology and created the foundation of the digital economy that transformed how individuals work, learn, and communicate.

The history of the computers could be traced through a number of different generations. The original generation (1940s–1950s) was based on vacuum tubes, where large, power-intensive, and high-priced machines with scientific and military applications, like ENIAC and UNIVAC I, were made. The second generation (1950s–1960s) brought about transistors and this led to smaller, faster, and reliable computers that used less power as an example the IBM 7090. The third generation (1960s–1970s) saw the introduction of integrated circuits (ICs), which made it possible to miniaturise, achieve more computing power and minicomputers such as the IBM System/360 emerged. The fourth generation (1970s–1980s) used the microprocessors and led to the personal computers like the Apple II and IBM PC. It was then succeeded by the era of early laptops (1980s) that despite being bulky brought about portability with the introduction of devices such as the Osborne 1 and the early Compaq models. By 1990s–2000s, laptop models got lighter, faster, and cheaper, becoming mainstream as a mid-range laptop. The 2010s began with the introduction of ultrabooks and 2-in-1 computers, which are ultra-thin computers with the functionality of a tablet-laptop, such as the MacBook Air and Microsoft Surface. With the new era of laptops (2010s–present), laptops have been made smooth and fast with SSDs, Wi-Fi, high-resolution screens, and extended batteries. The next-generation laptops (2020s and beyond) will be expected to incorporate AI, will have flexible displays, will be designed to be cloud-first and even have quantum computers. Figure 1.1 presents the history of computers and laptops across generations, with the machine of the 1940s vacuum tube to the current laptops and artificial intelligence-based models of the future.

1.6.4.2 Software Development for PCs

The emergence of personal computing resulted in a new era of software development, which is oriented to individual users and small businesses. This change resulted in the development of critical tools such as word processors, spreadsheet, and database management systems and it has radically changed productivity. The software industry thrived as the personal computers became easily available, providing applications that assisted in work, education, and entertainment. Mass-market programs and games were introduced to the houses all over the world, and the digital technology became an inseparable part of daily life. This software democratisation spread its wings and gave birth to innovation in user-friendly design and interactive computing experiences [11].

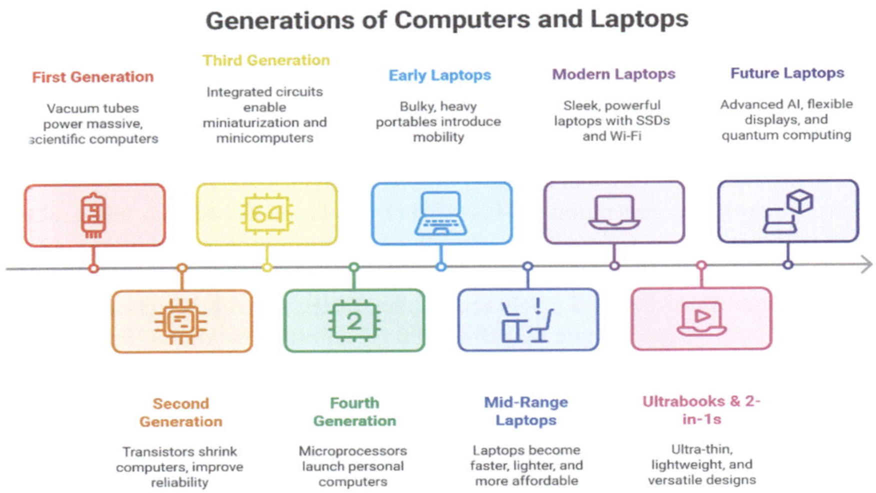

Fig. 1.1 Evolution of computers and laptops across generations

1.6.5 The Internet and the Modern Era

This computer science revolution did not only follow the expansion of the Internet but also followed the near-light-speed technological breakthroughs experienced in the late twentieth century and early twenty-first century. The future has continued to be re-moulded by new paradigms, technologies, and applications. The future has been characterised by new paradigms, technologies, and applications [12].

1.6.5.1 The Birth of the Internet

Internet has its roots in the late 1960s when the U.S Department of Defence created ARPANET—an experimental network to enable network communication to be done securely among institutions of research. This project brought the idea of packet switching, which involves the transmission of data in small manageable bits, and this greatly enhanced efficiency and reliability. ARPANET was the precursor to the interlinked networks that would later become the contemporary Internet and this was a breaking point in the history of digital communication.

1.6.5.2 From ARPANET to the Internet

The first long-range computer network in the world, which allowed communication between researchers by connecting the computers across a large distance, was a research project created at the United States Department of Defence in the late 90 s and named ARPANET; the network was based on a new technique known as the packet switching, or a technique of breaking up data into smaller packets to transmit it more efficiently. In the following 20 years, ARPANET was extended to move further and larger, with more protocols, such as TCP/IP.

For the next two decades, ARPANET continued to evolve further and bigger, with additional protocols, including TCP/IP. All these innovations have come to be widely used for transferring data on the Internet. These developments formed the base upon which the global network revolutionised communication and commerce.

1.6.5.3 The World Wide Web and the Digital Revolution

The World Wide Web came into being in 1991 by the British computer scientist Sir Tim Berners-Lee and it was a revolutionary system of interconnected documents in the form of hypertext that can be accessed through the Internet. This technology brought about a comprehensible world to everyone since users were able to navigate, access, and disseminate information and the world is informed in a democratic manner. The 1990s and early 2000s are commonly referred to as the boom years of the digital revolution, the boom of websites, the introduction of e-commerce, and the proliferation of online communication media. These advancements triggered the transformation of multimedia content and interactive digital environment.

Internet is a highly new technology that has been infiltrating life in a very deep and very far-reaching manner. Internet usage is now present in all spheres of activity, be it industry, education, healthcare, government, or culture. The modern world cannot imagine itself without referring to the Internet. Nowadays, information is being circulated all over in the universe. To make it possible for the web to reach consumers, businesses, and institutions, the World Wide Web is indispensable. These days, it's tricky to tell the distinction between the Internet and the global web.

1.7 Modern Computer Science

1.7.1 The Growth of Computer Science in the Twenty-First Century

Computer science has greatly evolved in the last century. As consequently, one of among the most notable fields that have impacted human history is computer science. Computer science is now at the forefront of innovation in every one of modern technological domains due to its rapid advancement and related fields like deep learning, big data, artificial intelligence, cloud computing, and cybersecurity. As a clue, computer science and related subfield technologies are essential to the contemporary healthcare, financial, cultural, educational, and governmental sectors, among other important sectors. By doing the provision of mobile devices, social media, and

Internet of Things-inspired control of traffic on roads, it has crept into daily human life. It has created avenues for knowledge and connectivity in a much more extensive way. Deep learning, self-learning systems, quantum information processing, and human–computer interactions represent some of the latest developments in computer science and technological advances. These are testing boundaries, shaping the technology's future, and having a previously unnoticed effect on society. The field's exponential growth has made computer science more than just a discipline. The discipline of computer science evolved into an interdisciplinary field of investigation that influences many other disciplines [12].

1.7.2 Artificial Intelligence and Machine Learning

The development of algorithms for machine learning (ML) and computational intelligence (also known as AI) is one of among the most recent and obvious revolutions in modern computer science. Artificial intelligence is the most essential element of advancements in computer vision, recognition of speech, processing of natural language, and machine learning for making decisions. Deep learning is contributing to considerable enhancements in artificial neural networks and other AI models. These approaches have produced developments such as personalised health services and autonomous vehicle technology that are revolutionising industries and reinventing the relationship between people and computers. With artificial intelligence and machine learning, the manner in which individuals used to interact with computers has completely changed from just a decade ago. The methods could help organisations to produce adaptable and useful insights while used alongside big data analytics [13, 14].

1.7.3 Data and Computational Infrastructures

With versatility, scalability, and novel application delivery methods, the use of cloud computing has boosted the capacity to create, store, and process huge quantities of data. The use of edge computing, which brings computation to the data sources for enabling real-time analytical tools with low latency constraints by autonomous or Internet-of-Things systems, has also grown into an equally important complement at the same time. In the age of the Internet, these networks are influencing how computation is distributed and utilised [15].

1.7.4 The Internet of Things and Ubiquitous Computing

Computer intelligence can be brought into the real world by means of the Internet of Things (IoT). Examples like smart homes, smart cities, intelligent medical equipment, and automated manufacturing show the full scope of the IoT's impact. There are many actuators and sensors that are used in the Internet-of-Things connected devices. The amount of knowledge flowing by means of these sensors and actuators is enormous. Numerous opportunities for innovation and decisions based on data optimisation are generated by the analysis of such enormous volumes of data. However, it additionally constitutes a serious problem for guaranteeing safety, safeguarding personal information, and seamless integration across heterogeneous systems [16].

1.7.5 Blockchain and Decentralised Systems

A decentralised system of secure exchanges has been rendered attainable by the use of blockchain technologies, which promote authenticity and reliability. Blockchain originally became well known owing to its utilisation in the market for digital currencies. These days, blockchain is used across numerous areas incorporating digital identity management, safe documentation, supply chain systems, etc. Besides, it has made transparent laser systems that are transparent, open, and immutable possible. In digital technologies, it has changed the traditional definitions of oversight, openness, and decentralisation. As a result, it has given novel solutions to handling reliability in the distributed network [17].

1.7.6 Cybersecurity in the Digital Age

As the digital infrastructures emerge in the online world, and it is yet to build a more comprehensive presence within that sphere, the digital assaults by malicious individuals are multiplying, becoming common and advanced. Cybersecurity has been a major issue that has seen its realisation to secure the data and systems in addition to the networks against more advanced risks. White-hat hacking and AI-assisted threat discovery is proving to be one of the key methods to punish the new threats and the constant dynamism of the innovation and vulnerability problem has described [18, 19].

1.7.7 Human–Computer Interaction and Robotics

As much as much of the computer science is the systems and data, there is always the human in which it plays a significant role. Technology allows the improvement of the human–computer interaction (HRI) so that it becomes usable and accessible. Synchronously in developments related to the sphere of robotics and automation, AI and mechanisation are linked and transform the industries, introducing autopiloted machines to the manufacturing industry, logistics, and healthcare [20, 21].

1.7.8 Quantum Computing and the Future of Computation

Beyond classical boundaries, quantum computing introduces novel principles, which were based on the qubits, superposition, and entanglement. It remains at the maturity stage but its cryptographic potential, optimisation, and discovery in science make it one of the most promising scanned computer science frontiers [22].

1.7.9 Immersive Technologies and the Metaverse

Finally, AI, IoT, cloud convergence, and immersive interface convergence are an indicator of the metaverse—a paradigm of convergence of physical and digital reality into interactive and enduring spaces. Through cloud, IoT, and computational intelligence, the world of the metaverse offers innovative technologies with immersive user experiences that change interactions with others, learning, and other workplaces. The metaverse also has problems and difficulties. In particular in the domains of digital wellness, morality, knowledge, and government. It involves implementing into practice an interdisciplinary scheme that can provide equal growth and sustainability [23, 24].

1.8 Ethical Considerations and the Future of Computer Science

Deep ethical questions which require careful investigation have been brought up by the development of information technology in time and its effects on other fields of work. Fairness, privacy, accountability, and the societal implications of emerging technologies are among the issues that are emerging in tandem with technological developments in the field.

1.8.1 Data Privacy and Security

Several computer-related uses collect, store, and interpret data. This has sparked worries about the confidentiality of users, and protecting it has grown to be a major part of the problem. The guidelines and rules to safeguard users' rights have been codified in the European Union's General Data Protection Regulation (GDPR). Data protection systems will be boosted by such legislative initiatives. Because of several shifts in a highly dynamic environment driven by newly emerging technologies, it is difficult and complex to ensure sufficient oversight on these regulations at any given time, particularly when new risks arise and pose serious moral information management issues. Such a dynamic and changing environment necessitates experts to discuss in detail their opinions regarding the importance of data protection, its nature, and the right techniques to be applied in this case. At the same time, the balance between computer science innovations and data privacy and protection should be considered. An environment of digitally globalised needs the presence of a responsible data management with appropriate ethics and technological capability.

1.8.2 Algorithmic Bias and Fairness

Bias is one of the most important and pressing ethical concerns in the algorithmic decision-making process. It is spatially very sensitive in the high-importance and highly sensitive contexts, such as law enforcement recruitment. The flaw of algorithmic bias and fairness is far more than a technical flaw, as it reflects and reinforces structural inequalities present in society. To remove these issues from the algorithmic decision-making, it requires the integration of different societal representative datasets. It also requires proper design principles, considering accountability and equity, specifically considering accountability and equity in the design framework. In human decision-making, artificial intelligence is overwhelmingly taking a key role. AI development must be focused and directed by ethical values so that transparency, fairness, and a responsible voice of society can be integrated.

1.8.3 Impact of Automation on Employment

Artificial Intelligence and robotics are having a big impact on automation. However, although the operational efficiency has improved due to automation in all industries of today, and as we are going from Industry 4.0 to Industry 5.0, this operational efficiency will definitely increase. But the concern about employment is rising in this scenario. The cost reduction has been significantly reduced due to automation, which is involved in production, distribution, and service delivery. But this shift is adding a set of challenges to society. The first challenge is the substitution of traditional jobs and the nature of work. This is creating a revised labour market. Many of the traditional labourers are getting out of their jobs, and creating unemployment. It has to be taken care that automation is supported with the rescaling and upscaling of the workers. Otherwise, the digital economy created based on automation will generate a lot of unemployment. We must make necessary arrangements to ensure the workforce is prepared for the challenging market and that employment becomes sustainable and inclusive.

1.9 Roadmap of the Book

The main innovations, approaches, and algorithms influencing the modern digital world are explored in this book, which has been composed with the reader in mind. While every chapter exists on its own, they operate together to seamlessly lead readers from basic computer science concepts to innovative developments that are revolutionising society. Thirteen chapters, each devoted to a significant growing theme, constitute Sect. 1.7, which emphasises rapidly growing areas of computing today. The roadmap assists readers in understanding the broader context regarding information science's digital evolution by outlining the connections between each of these concepts. Academic rigour and practical simplicity are balanced in the language.

The first chapter, Generative AI and Creative Content Generation (Chap. 2), broadens the scope of creativity while posing ethical and ownership concerns by demonstrating how machines produce text, images, and music. The introduction of quantum computing is then covered in Chap. 3, along with how its tenets, techniques, and paradigms may redefine computation and have an effect on large-scale optimisation and cryptography.

Chapter 4 describes the data analytics techniques of computer science. Knowledge from the raw data is usually retrieved using a broad set of machine learning techniques and algorithms. Thus, this chapter is named as Data Analytics and Machine Learning.

As AI technologies advance, so too do the challenges they raise. Artificial Intelligence and Ethics (Chap. 5) addresses fairness, accountability, and data privacy issues. It acts as a counterbalance, ensuring that the technical achievements of preceding chapters are viewed through a human-centred and responsible lens.

From there, we explore the infrastructures enabling intelligent systems. Big Data and Data Science (Chap. 6) examines the processing and visualisation of large-scale datasets, while Blockchain and Cryptocurrencies (Chap. 7) introduces decentralised trust mechanisms that underpin secure transactions and novel digital economies.

Building on the security theme, Cybersecurity and Ethical Hacking (Chap. 8) provides insight into safeguarding digital ecosystems. The tools, vulnerabilities, and defensive strategies discussed here are essential for sustaining trust in a hyperconnected world.

The book then broadens into the physical–digital interface with Internet of Things (Chap. 9), which explores how connected devices generate data and enable

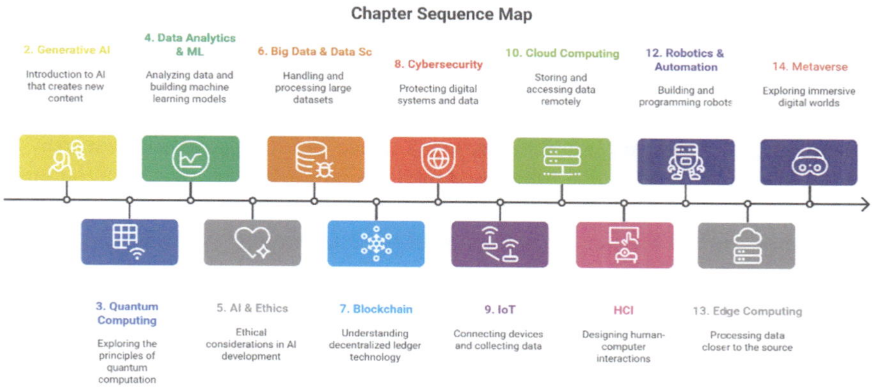

Fig. 1.2 Chapter sequence map illustrating the progression of topics in modern computer science

innovative environments. Complementing this is Cloud Computing and Virtualisation (Chap. 10), which illustrates how scalable infrastructures support both IoT deployments and enterprise applications.

The human dimension is revisited in Human–Computer Interaction (Chap. 11), emphasising usability and accessibility as critical factors in technology adoption. From here, Robotics and Automation (Chap. 12) demonstrates how AI and mechanical systems converge to augment manufacturing and healthcare industries.

Recognising the need for speed and proximity in computation, Edge Computing (Chap. 13) explains how data processing closer to the source reduces latency and empowers real-time decision-making. This naturally sets the stage for the concluding vision in The Metaverse Revolution (Chap. 14), which examines immersive digital worlds reshaping work, learning, and social interaction.

These chapters provide readers with a panoramic view and detailed insights into the technologies that are converging and transforming modern society. The structure of the book, outlined in Fig. 1.2, highlights how each chapter builds on the previous to provide a comprehensive understanding of contemporary computing technologies.

1.10 Summary and Conclusion

Computer science has evolved from abstract mathematical reasoning to a dynamic and interdisciplinary domain that powers nearly every facet of modern life. This chapter has provided an overview of its scope, history, and significance, from early computational theories and pioneers such as Ada Lovelace and Alan Turing to landmark innovations like the ENIAC, personal computers, and the Internet.

The development of high-level programming languages has had a big impact on the development of computer science applications. High-level programming languages provide abstraction in terms of syntax and structures, allowing developers to focus on the application logic without worrying about machine-specific requirements. This has improved the productivity of the programmers. The innovation of the personal computer and the Internet has transformed computer science into an integral part of daily life. Computer science and its applications can now be used by ordinary people with ease. Machine learning, cybersecurity, artificial intelligence, blockchain, quantum computing, robotics, human–computer interactions, and immersive technologies are just a few of the high-tech fields it encompasses.

The majority of other fields of work, including education, industry, communication, and entertainment, are being revolutionised by these branches of computer science. In the meantime, it has brought up some ethical concerns about data security and privacy, as well as equity and unemployment because of automation. This chapter will offer a framework for additional research as it examines the historical underpinnings and prospects of computer science. It is also true that computer science should be seen as a tool for social change as well as for technological advancement. It will motivate readers to act responsibly in response to these opportunities and challenges.

References

1. Fu Z, Cao Y, Zhao Y. Identifying knowledge evolution in computer science from the perspective of academic genealogy. J Informet. 2024;18(2): 101523.
2. Toukiloglou P, Xinogalos S. Effects of collaborative support on learning in serious games for programming. J Educ Comput Res. 2025;63(1):126–46.
3. Sardar TH. Reflecting on a decade of evolution: mapreduce-based advances in partitioning-based, hierarchical-based, and density-based clustering (2013–2023). Wiley Interdiscip Rev: Data Min Knowl Discov. 2024;14(6): e1566.
4. Prasanna VL, Rao SG, Vaibhav KS, Shekhar KC. The nexus of innovation: visionaries shaping the past, present, and future of technology. In: Multi-disciplinary research and sustainable development, p. 364–374. CRC Press.
5. van der Waerden BL. Three Muslimic Authors. In: A history of algebra: from al-Khwārizmī to Emmy Noether. Berlin, Heidelberg: Springer Berlin Heidelberg;1985. p. 3–31.
6. Jesiek BK. The origins and early history of computer engineering in the United States. IEEE Ann Hist Comput. 2013;35(3):6–18.
7. Aikat DD. Pioneers of the early digital era: innovative ideas that shaped computing in 1833–1945. Convergence. 2001;7(4):52–81.
8. Kostitsyna I, Wood C, Woods D. Turning machines. In: 26th International conference on DNA computing and molecular programming (DNA 26)(2020). Schloss Dagstuhl–Leibniz-Zentrum für Informatik;2020. p. 11–1.
9. Randell B, editor. The origins of digital computers: selected papers. Springer;2013 Dec 21.
10. Waldrop MM. The origins of personal computing. Sci Am. 2001;285(6):84–91.
11. Alajrami S, Romanovsky A, Gallina B. Software development in the post-PC era: towards software development as a service. In: International conference on product-focused software process improvement. Cham: Springer International Publishing;2016. p. 662–671.
12. Miao J, Wang Z, Wu Z, Ning X, Tiwari P. A blockchain-enabled privacy-preserving authentication management protocol for internet of medical things. Expert Syst Appl. 2024;1(237): 121329.
13. Hassan MA, Sardar TH, Fahim MF, Mohamed MS, Suleyman RM, Usman MM. Artificial intelligence and deep learning applications on big data computing frameworks. In: Big data computing. CRC Press;2024. p. 196–211.

14. Sardar TH, Pandey BK, editors. Machine learning hybridization and optimization for intelligent applications. CRC Press;2024.
15. Barua R, Hazarika H, Jena JJ, Jha AV, Sardar TH, Gourisaria MK. Leveraging convolutional neural networks for accurate skin cancer classification. In: 2025 5th International conference on expert clouds and applications (ICOECA);2025. p. 572–577. IEEE.
16. Usman MM, Fahim MF, Al Mahedi Hassan M, Sardar TH, Adhoni ZA, Gourisaria MK. Harnessing IoT and machine learning for sustainable, smart urban environments. In: Applications of computational learning and IoT in smart road transportation system. Cham: Springer Nature Switzerland;2025. p. 199–214.
17. Sardar TH, Dar SA, Indhumathi R, Dhasmana G, Prasad G, Kumar P. Integrating blockchain and quantum key exchange with deep learning for enhanced medical data. Procedia Comput Sci. 2025;1(259):1208–17.
18. Kumar C, Kumar KA, Gupta S, Sardar TH. Cyberbullying detection based on the fusion of DistilBERT and SIMHASH Technique. In: 2024 2nd International conference on artificial intelligence and machine learning applications theme: healthcare and internet of things (AIMLA);2024. p. 1–4. IEEE.
19. Rizvi I, Raj S, Singh V. Cybersecurity in the digital age. In: Technology for societal transformation: exploring the intersection of information technology and societal development. Singapore: Springer Nature Singapore;2025. p. 131–148.
20. Ho MT, Vuong QH. Five premises to understand human–computer interactions as AI is changing the world. AI & Soc. 2025;40(2):1161–2.
21. Firoozi R, Tucker J, Tian S, Majumdar A, Sun J, Liu W, Zhu Y, Song S, Kapoor A, Hausman K, Ichter B. Foundation models in robotics: applications, challenges, and the future. Int J Robot Res. 2025;44(5):701–39.
22. Alrashed S, Min-Allah N. Quantum computing research in medical sciences. Inform Med Unlocked. 2025;1(52): 101606.
23. Singh R, Sarkar P, Bandyopadhyay A, Swain S, Sardar TH, Pandey B. Combinatorial kinetic dualistic auction model for metaverse service allocation. In: 2024 IEEE AITU: digital generation;2024. p. 27–32. IEEE.
24. Murala DK, Panda SK. Metaverse: a study on immersive technologies. Metaverse Immersive Technol: Introd Ind, Bus Soc Appl. 2023;20:1–41.

Chapter 2
The Rise of Generative AI: From Foundations to Creative Frontiers

2.1 Understanding Generative Models

The generative AI is a new concept that has come out from the traditional artificial intelligence systems. It is actually a benchmark breakthrough of general artificial intelligence. In the case of generative AI, we can actually generate new content. Unlike the traditional artificial intelligence system, which can actually be used for detecting certain objects by creating boundaries in the particular data and learning new patterns. The generative AI can read through the training data, understand the underlying structures and distribution, and create new content. For example, imagine that you have been provided an image and you have been told to distinguish between the cats and dogs available in that image. The generative AI systems, unlike traditional AI, which can ONLY identify cats and dogs in the image, can actually comprehend the entire scenario, and based on your request, it can generate cats and dogs or some other objects and put them inside the image as if it were a real one. We should not mix generative AI with the discriminative models. In discriminative models, we can understand the data, and we can draw the boundaries between different objects and categories. In this way, discriminative models can understand the data patterns but cannot generate new objects or patterns and add to them. But in generative models, we can understand the full landscape of the data using the data training and enable it by adding novel data input to it.

Generative AI systems are based on generative models. These models are versatile in nature. For example, it can be used for creating almost anything required for modern times. We can create text, we can create chapters, we can write books, we can write different kinds of documents, and we can even create experimental results using it. We can use these generating models for generating new images, modifying existing images, adding new features to the images, etc., as well. Similarly, we can even generate new videos which were not available before by using these generating models. Even though there are many chatbot systems available based on generative models. This can be used across different industries, such as creating art, music,

T. H. Sardar, *The Spectrum of Computer Science: Emerging Technologies and Trends*, Studies in Smart Technologies, https://doi.org/10.1007/978-3-032-19632-3_2

drawing, paintings, making video graphics, creating teaching content, and even for improving business processes.

In this way, generative AI models are doing what the general artificial intelligence models cannot do. However, the generative AI models are based on two practical and already proven architectures. These architectures are known as generative adversarial networks (GANs) and variational autoencoders (VAEs).

2.1.1 *Generative Adversarial Networks (GANs): a Revolutionary Breakthrough*

The GANs were introduced by researchers in 2014. Thus, this concept is a relatively recent concept proposed by Ian Goodfellow [1]. When he first produced this idea, it caught the eyes of the researchers and became a breakthrough in machine learning over time. GANs are based on two concepts: one is the generator and the other is the discriminator. For simplicity, you can consider the generator as an artist who is going to create new art and the discriminator as a critic who is always criticising the newly created art. The generator can act as an artist as much as it has been trained by the training data. It is a very important factor in how the generator has been trained with the data. Generally, the generator is trained with a huge volume of data. Based on that training dataset, it can understand the pattern and generate new art. The discriminator will act as a critic of the artist. It also depends on how much it sees the newly generated art as perfect or not. If it finds that the newly generated art is not correct, or it is fake, or it is misplaced, then it will criticise the generator. In this way, both of them balance and complement each other. Simplifying this, we can consider that the generator is an artist who is generating art, and the discriminator is a teacher who is criticising that art. The criticism which comes from the teacher will provide positive feedback to the artist to create an improvised and realistic version of the work. The generator and discriminators work in parallel, and both are run by neural network systems. They complement each other. This process produces new data, whether it is image, audio, video, or text.

The competition between the generator and the discriminator makes both of them powerful. They complement each other to improve both of them. The generator will become more skilled at producing realistic, non-fake, and accurate outputs, while the discriminator becomes more accurate in detecting the fake and non-realistic things. The training keeps on going unless it reaches a certain point of balance. This point of balance is known as the Nash equilibrium. The generated creation becomes inseparable from the real-world input data to the GANs.

However, GANs face many challenges. Training can be unstable, and issues like mode collapse (where the generator produces limited varieties of outputs) or vanishing gradients (which slow learning) can arise. Careful tuning of hyperparameters and network architectures is essential to maintain stable and effective training.

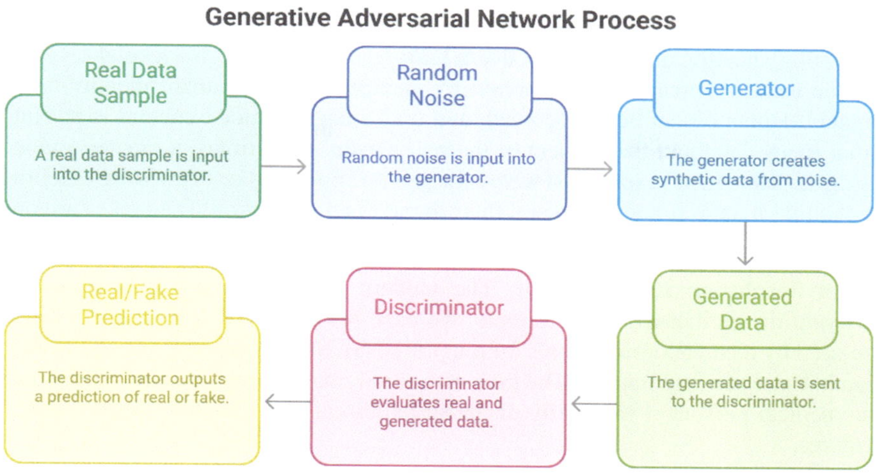

Fig. 2.1 Overview of the generative adversarial network workflow

Example: A GAN trained on fashion images can generate entirely new clothing designs that are coherent in style and pattern, giving designers inspiration without copying any existing piece. Similarly, in medical imaging, GANs can create synthetic but realistic scans to augment datasets for training diagnostic models.

Figure 2.1 depicts the Generative Adversarial Network (GAN) process, where random noise is input into a generator to create synthetic data, which is then evaluated by a discriminator alongside real data samples to predict real or fake outcomes.

2.1.1.1 Technical Applications of GANs

Generative Adversarial Networks (GANs) have proven to be remarkably versatile, influencing both research and real-world practice across numerous domains [2, 3].

Image Generation and Synthesis: GANs are widely used to create photorealistic images that maintain intricate detail and coherence. For example, the fashion industry applies GANs for virtual clothing prototypes and digital fashion shows, while architects use them to visualise building concepts before construction. There are many tools which have been developed by large enterprises for creating their own version of GANs. Nvidia has its own GAN, which is named StyleGAN. StyleGAN has created high-resolution faces of humans and similar animals, which look very original, and it cannot be easily known as a deepfake.

Deepfake Technology: As we know, the GANs can be used for learning through image and video datasets as well. It can learn from human faces, and that can be used to generate human faces, but in different contexts. Such a kind of application of GANs can be very effectively used in application areas such as movie creation and educational video creation. However, the side effects and the bad use of such technology have created a major concern known as “Deepfake”. In Deepfake, GANs can be used to spread misinformation by manipulating the human identity-based content. Human identities are copied, and their image of facial content is put into other images without the consent of the real person. Sometimes, it creates a video using the same facial expression where the person appears to be expressing or telling something inappropriate or beyond their dignity and reputation. Deepfakes are one of the biggest ethical concerns of GANs.

Super-Resolution Enhancement: The modern medical diagnosis is based on different image datasets. For example, we have MRI and CT scan images, which are used by medical professionals for diagnosis. GANs can help improve the resolution of these medical images. The high-resolution images provide better clarity, and the medical personnel can easily diagnose the disease by obtaining better clinical features.

Text-to-Image Generation: Modern GAN architectures are capable of understanding natural language text, and based on the provided details, they can create images. The better the description of the text is, the better context-aware images can be created by the GAN. This particular application of GANs can be used fully for educational purposes, for simulation purposes, and for creating novel and Creative arts.

AI-Generated Artworks: GANs are also capable of generating AI-generated artworks. The generator artworks are so real that it has now been accepted by a portion of the mainstream market. A famous example is “Edmond de Belamy”. This GAN-generated image was sold for 432,500 dollars. This kind of production of art has created a debate about whether the ownership of the particular image belongs to the AI or the person who uses the AI.

Artistic and Cultural Synthesis: GANs can also be used to mix cross-cultural art forms, different region-specific styles, geographic-specific influences, and society-specific cultures, etc., together to create a hybrid art. In this way, GANs can create the highest form of human creativity and collaboration.

2.1.2 *Variational Autoencoders (VAEs): A Probabilistic Approach to Generation*

The Variational Autoencoders (VAEs) were actually introduced before generative adversarial networks. The generative adversarial networks were introduced in 2014, but the variational autoencoders were introduced in 2013 by Kingma and Welling. It’s a significant advancement in generative modelling. It is based on mathematical principles and especially on probabilistic theorems. VAEs rely on basic inference and variational optimisation theorems [4]. VAEs are more specifically designed for those applications which require accurate generation of content with good interpretability and smoothness during object creation.

The variational encoders are formed by two components: one is the encoder and the other is the decoder. The encoder will compress the complex input data into a lower-dimensional space, and it will be represented in a simple way in a latent space. The decoder will construct the data by creating new samples from the space created by the encoder. The encoder created a latent space that is structured and continuous. As it is structured and continuous, if we have any change in it, even a tiny one, it will correspond to a big change in the output. For example, consider that the encoder has created a latent space based on a human face. Now, if we go through the latent space and trend, do the transition, then we can see the face changes from smiling to neutral or from a big smile to a small smile, etc.

In this way, VAEs provide a framework for controlled generation of synthetic data, simulation, and also for having a new set of creativity. VAEs are also designed to deal with uncertainty better than GANs. These features of VAEs make it offer better confidence in the generated generative AI output. A big application of VAEs is currently happening in the healthcare sector. For example, sometimes we are required to generate a synthetic artificial biomedical dataset because we cannot provide real data without patient consent. Such an artificial biomedical dataset can be easily created by VAEs and used for research and development.

2.1.3 Emerging Generative Models

Generative AI is a continuously developing field. It is not confined to the GANs and VAEs only. There are many emerging models which are creating state-of-the-art capabilities and producing content like never before. This subsection will discuss these emerging models.

Diffusion Models: The New Paradigm in Image Generation

The diffusion models have created a significant difference in text-to-image generation by their unique approach of denoising. Like the GANs, which use adversarial processes, diffusion models transform random noise into structured data. This conversion happens using an iterative denoising process, which is built into diffusion models [5]. The diffusion model's strength lies in its data training process. During the data training process, they are equipped to learn using a reverse noise addition process. This noise adding process adds extra strength to the diffusion model as its output becomes more trustworthy, orderly, and of better quality. There are many diffusion models available in the market today, but the most popular models are Stable Diffusion, DALL.E 2, and Midjourney. These diffusion models can create high-resolution multiple versions of images from natural language textual inputs. They are, in general, known as outperforming GAN models by providing a better quality set of images with a good context and better apprehension of diversity. For example, if an artist generates an image using a diffusion model by giving a general natural language input, a garden in heaven with multiple trees, then the user will get similar variations of gardens with different visuals inside a heavenly background [6].

Transformer-Based Generative Models: Language and Beyond

The transformer-based generative AI models have created a big breakthrough as they can support multimodal data to work efficiently with text, images, and other types of data. It can deal efficiently with advanced reasoning and create writing abilities to the next level. It can also provide problem-solving capabilities. GPT series, like GPT 1 to GPT 4, are a live example of the practicality and wide popularity of transformer-based generative models. There are some cross-modal transformer-based generative AI models which can even be used for interpreting language and image data together [7].

Multimodal Models: Unified Content Generation

Multimodal models have created a big breakthrough in generative AI ecosystems. They are used for unified content generation based on different sets of data inputs, such as text, audio, video, and even programming code. They can perform this data processing at the same time, simultaneously. For example, ChatGPT provides reasoning along with the images provided, tied to different kinds of human languages. Another example of Google Gemini can also be given a nomination. It is designed as a multimodal unified content generation GPT. At the same time, Gemini can generate text data while interpreting a visual interpretation of an image at the same time. It can provide the text with a consistent programming code which is interpretable and compiled. It can even specify the online URL to compile the code, how to compile the code, and how to create a project out of it. Tools are created as the next step towards human–computer interactions, which is to engage like it works with her, neither human nor an expert [8].

Flow-Based Models: Precision and Probabilistic Control

The precision of the produced output is always a concern for the generative AI. The flow-based models are designed to address this particular concern. Flow-based models are particularly good at finding similar data points with high precision. This helps it achieve high precision of the generated output. This is achieved technically by transformation between the real-world complex distribution of data objects in a latent space with respect to simple probability distributions [9]. The flow-based models are highly appropriate and useful for those scientific domains where probabilistic reasoning is important and noise in the input dataset is usually very high.

Convergence and Future Directions

The current development of generative AI is now moving towards a convergence so that the generative content becomes more accurate. For example, these new generative AI applications are using diffuser–transformer hybrids for better quality of image generation, while using GANs and VAEs for balancing diversity and to gain more control in content generation. This kind of convergence-based modern generative AI is more powerful than the traditional ones, and they also provide stability, customisability, and are ethically sounder.

Technical Implementation Considerations

Many technical implementation considerations have to be taken into care when you want to deploy a generative AI system [10].

Hardware and Cloud Computing: The generative AI models are very much computing-intensive. Having said that, we can understand that generating equality generative AI requires more powerful processing systems than the central processing unit that are built into commodity computing systems. We also need Graphical Processing Units (GPUs) and Neural Processing Units (NPUs) to efficiently use these models. Modern cloud platforms are providing services based on chargeable subscription requests. These cloud service providers are smarter in providing these services by sharing the high-end processing power installed at their server level. The developers do not need to pay for the hardware directly for their implementation, but rather just rent it from a cloud service provider. Examples of this cloud service provider are Google Cloud, Microsoft Azure, and AWS. This platform is allowing different developers and companies to innovate modern generative AI content without bearing the direct expenses of the hardware components.

Model Training Best Practices: Generative AI models are very much dependent on the quality of their training. The training requires the use of diverse noise-free datasets. The data preprocessing also takes a major role in determining the quality of generative applications. One of the best practices of model training comes into the picture. Learning is being used by the expert so that they can use find-tune and domain-specific datasets to gain strong performances.

Fine-tuning and Customisation: Some of the applications require fine-tuning and customisation for generating quality domain-specific generative AI content. There are some techniques developed to attain this objective. For example, Low-Rank Adaptation (LoRA) is developed for a large model to get trained for specific domain outputs. Retrieval Augmented Generation (RAG) is also used for developing highly efficient output by enabling generative models to fetch external knowledge, as the generative AI model produces results.

2.2 Applications of Generative AI

2.2.1 Broader Applications of Generative AI

The generative AI applications are penetrating every sector of work [11]. All the applications of generative AI are now brought here. This attempt is to showcase the key application domains of generative AI.

Drug Discovery and Molecular Design

In the traditional pharmaceutical domain, drug synthesis research is a time and cost-intensive process. Generative AI solutions are being used by the pharmaceutical companies to shorten the time and lower the cost. Generative AI applications can design novel drug molecules which are beneficial for a certain drug composition; at the same time, they can discover the toxic elements in the drug, and their side effects are also discoverable now. Using these generative AI applications, we can propose new chemical component structures, determine their stability, and have a probabilistic base for drug target achievements. For example, one pharmaceutical initiative named "Silico Medicine" has already used generative AI to generate the composition of a medicine.

Code Generation and Software Development: Modern generative AI applications can generate programming language code based on human input of natural language text descriptions. It is now even possible to provide a specific programming language-specific code from the text description. One step ahead, now these generative AI applications can help debug the code and suggest suitable online or offline applications to compile the code and collaborate with other programmers. These tools are helping programmers to develop better quality code and increase their productivity. It even enables naive programmers to develop advanced code-based applications, which were previously done by expert programmers only.

Data Synthesis and Augmentation: It is well known that for a machine learning algorithm to work effectively, a proper dataset is required. But the real-world datasets are costly. At times, the dataset is difficult to retrieve due to ethical complexities or issues. For example, some of the medical datasets are very difficult to obtain, as a dataset cannot be created without the consent of a patient. Generative AI applications are helping researchers in this segment as well. Now, generative AI applications can create synthetic datasets while preserving their statistical properties as if the dataset were an original one. Status set, diesel artificial, it's the feature of a real-world dataset and can be used ethically by researchers.

3D Content Creation and Design: Content generation is very much required for a few sectors. For example, the field of architectural designs, manufacturing centre planning, to design modern cities, and for modern computer games. 3D content, especially images and video generation, has become very crucial. Generative AI is now enabled to create 3D models. Models can effectively design different kinds of 3D diagrams and videos. These AI models can be used by architects for city planning, building planning, and manufacturing centre planning. At the same time, video game developers can use this for developing immersive video games.

Natural Language Processing: Generative AI applications are adapted for answering users' natural language queries. Generative applications can work as a dialogue generator, as if the user is chatting to a real-world human being. It can translate one natural language into another. It can summarise a large amount of content into a smaller amount of content. For example, ChatGPT can provide the abstract of a large research paper. These intelligent tools are being deployed by companies for customer service, providing content, in the publishing process, etc.

Financial Modelling and Analysis: Generative AI applications can now be used for financial modelling and analysis. Generative AI applications can generate synthetic financial data. This data can be analysed to check the risk of the financial market for a particular scenario. It can also be used for generating market conditions and help in understanding which risk-taking is positive for a certain condition, and to understand the stress behind loans and other capacities of a financial firm. In this way, a generative AI application can provide market strategies and decision-making to the financial companies.

Scientific Research Acceleration: Generative AI can work as a catalyst for scientific research and development. Using generative AI, the scientist can now generate a synthetic artificial dataset which has similar statistical properties to a real-world dataset. It can be used for experimentation, and it's ethical. The scientist can also write their research paper with the help of the generative AI applications. It is also ethical to generate the structure of a scientific paper using generative AI applications.

2.2.2 Creative and Cultural Applications

The generative AI applications are not only limited to the technical field but also encompass creativity and cultural applications. Some of these applications are provided below:

Digital Art and Visual Expression: There are many generative applications which can be used to create new visual scenarios for digital art and visual expression. Generative AI applications can be used to create film-like environments without going to a particular physical place. Also, these applications can be used to reimagine existing or old scenarios and to blend them with the modern environment.

Musical Composition and Sound Design: Generative AI applications can single-handedly generate musical compositions. These musical compositions appear to be very real and consist of intelligent mixing of multiple musical instruments. To the next level, generative AI applications can mimic the voice of a singer based on their existing recordings. A new song can be sung by an old artist with the help of these applications. In this way, generative AI can also provide a personalised playlist based on a person's personal preferences over time.

Literary Arts and Narrative Generation: During the initial days of generative AI, its capacity was limited to generating short paragraphs. It is now capable of generating a full-length story with a practical plot, and the characters can have a proper role in it. Modern generative AIs are even capable of generating interactive novels, series of stories, and collaborative writing and many other Creative arts.

Interactive Storytelling and Immersive Narratives: Interactive storytelling is a new entertainment method where the user chooses the flow of a story. This kind of story is also known as adaptive stories. Generative AI is now capable of generating such adaptive stories. Adaptive stories are widely used in video games, in medical treatment programs, and in educational simulation programs. The users can personally engage in the storyline as it provides an engaging experience based on their own wishes rather than the description of a storyteller.

Gaming and Virtual Character Development: In video games, there is a concept called non-player characters (NPCs). NPCs are not controlled by the video game player, but it is managed by the artificial intelligence that runs the game. Advanced systems are changing the way non-player characters are designed. In the enhanced generative AI-based games, NPCs are now providing contextually relevant, engaging textual expressions, adapting to the players' behaviour, etc. These NPCs can also remember the last stages of the game and play accordingly. This has created a video game that comes in a realistic and immersive way. This further enhances the engagement of the real player in the game.

Educational Content and Training Applications: In the traditional teaching learning process, the teacher teaches and the student listens and learn from the lectures. In the generative AI-based teaching–learning process, the teaching content can be created to best meet the students' needs. An active teaching process can help students understand the knowledge through interactive quizzes, customised case studies, and interactive tutorials. Professionals can understand a particular work by understanding its overall implication in the large scenario of the domain.

Figure 2.2 shows generative AI application distribution across key sectors. We can see that, from Fig. 2.2, scientific research has a 25% contribution, similarly, creative industries and commercial applications have 35% and 40% contributions, respectively.

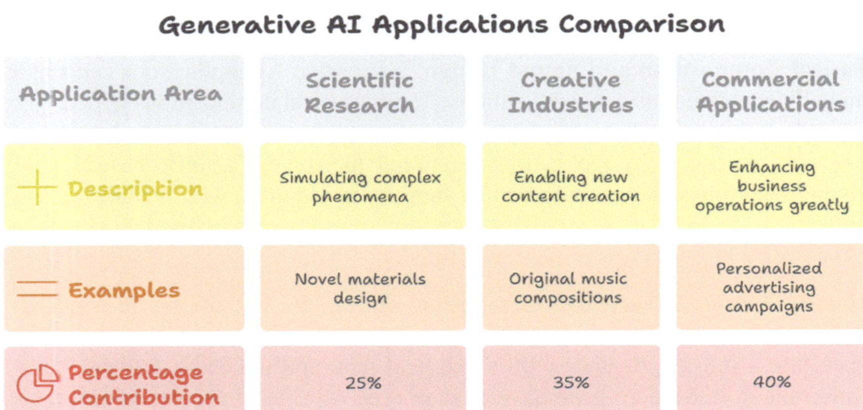

Fig. 2.2 Generative AI application distribution across key sectors

2.3 Challenges and Ethical Considerations

2.3.1 Ethical Concerns in Content Generation

2.3.1.1 Deepfakes and Misinformation Proliferation

The biggest and most widespread misuse of generative AI is Deepfake. In Deepfakes, the generative AI is used to generate identity-based content without the permission of the person who is being shown in the Deepfake images or videos. All the same applications of generative AI can be used positively for the movie industry, the educational sector, and for historic narrative building. But in Deepfakes, it is misused for putting such words in people's mouths which they have never said, or to generate adult nude content for innocent persons' bodies, which becomes a big concern. Sometimes, Deepfakes are widely used for online harassment, for making financial and criminal fraud, and for spreading political propaganda [12].

The consequences are not limited to individuals; they extend to entire societies. Deepfake-driven disinformation campaigns can undermine democratic elections, erode judicial credibility, and weaken social trust. When people can no longer reliably distinguish truth from fabrication, the very idea of shared reality is put at risk.

Governments, researchers, and technology platforms are working on countermeasures such as detection algorithms, criminal penalties for malicious use, and stricter content policies. However, the pace of generative AI innovation often outstrips these safeguards, making this an evolving cat-and-mouse problem.

2.3.1.2 Intellectual Property and Creative Ownership

Another pressing concern is intellectual property (IP). Traditional IP laws assume human authorship, but AI blurs this assumption by generating music, art, literature, and even inventions [13]. If a fashion designer co-creates a collection using AI-generated designs, who owns the rights—the human, the AI's developer, or both? Similarly, when large datasets are used to train generative models, copyright questions arise: should the original creators be compensated?

Current legal systems provide few clear answers, and different countries are adopting divergent approaches. In the United States, the AI alone cannot hold any copyright. In different parts of the world, we can see an inclination towards the hybrid approach. It is a clear sign of the time to update the international agreements and their rules and regulations, such that human creativity can be encouraged through loss, while it can use fairly generative systems.

2.3.2 *Technical Barriers and Limitations*

2.3.2.1 Content Quality and Algorithmic Bias

The content quality of a generative AI actually lies in its training process. While training, the provided data can have certain biases: they may be based on cultural biases, gender biases, religion-based biases, or regional biases. These biases are reflected in the generative AI output of texts, audio, videos, and images. It also sometimes creates inconsistent logic; sometimes its output looks fancy, but actually hallucinates with the weaker output [14].

As a solution, the training dataset should be free from biases. Diversified sources should be used for selecting the training datasets. We can also include some techniques and methods to remove the bias from the dataset. Sometimes the shuttle cultural basis is not easy to detect; in some cases, we should employ expert teams to identify and remove such hidden biases.

2.3.2.2 Resource Intensity and Accessibility Challenges

The generative models required a lot of energy and computing power to work; some of the genetic AI models require millions of GPU hours. With the GPU, it now requires Neural Processing Units (NPUs) hours as well. The question of all these processing systems also requires a huge amount of energy [15]. Also, it requires a specialised expert to be involved in the generative AI projects. Create issues of environmental sustainability and accessibility challenges. The huge processors and computing require energy that can affect the carbon footprint, which in turn damages global sustainability goals. Regarding the accessibility challenges, only a big organisation with a huge amount of money invested through shares or by its stakeholders can run such a generative AI organisation. As a solution, we developed efficient architectures which can consume less energy but provide a good amount of computing power, and the system should be developed in such a way that general small companies and developers can collaborate to create standard generative AI models.

2.3.3 *Responsible AI Development and Future Directions*

2.3.3.1 Establishing Ethical AI Frameworks

The technologists and policymakers should have a good collaboration with the Atheists and civil society members so that they can guide the ethical AI frame of development [16]. Establishing ethical AI is always challenging work. Because establishing this kind of framework requires active participation from atheists and civil society members.

We should ensure that the data training happens only through unbiased and ethical diverse datasets. Has to be noted that the limitations of the AI systems should be clearly stated to the users. The genetic framework is supposed to assimilate, understand, and reflect the cultural differences and ethical values it carries.

2.3.3.2 Emerging Opportunities and Future Innovations

As the generative AI faces many problems, for example, Deepfake and hallucination problems, there lie huge possibilities for development in the future as well [17]. If the generative AI models become reliable, it is free from hallucinations and can give controlled, ethical, and technically correct outputs while providing fast responses, which would be a great advantage for humanity. For example, we have a new concept known as explainable AI (XAI). Explainable AI is a technique which helps users understand how the model processes certain input and produces the output. In this way, XAI increases trust in an AI system.

However, the most promising future of innovations in generative AI lies in human–AI collaboration. AI systems in this scenario will not replace humans but will collaborate with humans so that they can enhance human abilities and create next-level content.

2.4 Industry Case Studies

Film and Entertainment: The generative AI is now being used practically in the entertainment industry. As an industry case study, we can name Cinelytic's AI collaboration with the famous movie-making enterprise, Warner Bros. Their collaboration can get a better marketplace for their plots, they can evaluate based on their economic viability, based on the audience preferences, and how a particular movie can attract the maximum audience. AI is not working here as a storyteller but as an advisor.

Fashion and Retail: Similarly, in the fashion industry, generative AI has created an impact. As an example of an industry case study, we can name Stitch Fix's employment of generative AI. That generative AI recommends dresses based on personal preferences to its customers. Based on the season, the quality of the fabrics, the previous experience of the customers, and their earlier dress choices. This way, generative AI recommends personalised preferences for dresses, which is a big shift from a one-size-fits-all approach.

Journalism and Publishing: Sometimes, generative AI is used for a certain cause in journalism. For example, the Washington Post uses a Heliograf AI so that their large news articles can be prescribed to the readers as abstracts or summaries in the first place. Similarly, some general journalists also use generative AI so that their deeper investigations can help out, while opening the concepts and their interconnections, organised knowledge derivation becomes easier from the generative AI systems.

Advertising and Marketing: A leading example of advertising and marketing, and generative AI is produced by Ogilvy. Ogilvy can quickly understand the demographic and region-based requirements for a certain advertisement and marketing campaign, and customise its advertisement based on the scenario, while maintaining the creative quality and understanding the ground realities. It produces multiple variations of the same adverts in a much shorter amount of time than the traditional systems.

2.5 Future Technological Directions

The evolution of generic AI is happening in a very random and fast way. Its direction is mainly aligned with three critical ways:

1. **Enhanced Multimodal Capabilities**: Although existing generative AI models have capabilities to generate multimodal content, in future, these multimodal capabilities will be enhanced in a totally different and very high-efficiency way. For example, the future of future generative AI will be able to create videos with images, audio, text, and stories altogether while providing the plot and their character, making their own voice records, and producing a soundtrack which can run in the background. All will work in harmony, and it will give an exact, real-world feeling to the audience.
2. **Improved Efficiency and Accessibility**: There is intense research happening on generative AI, such that it can go for learner architecture and the training strategies get into novel terms, so that the computational cost can be reduced. Due to computational costs, generative AI systems cannot be maintained or created by ordinary organisations and general developers. When these improvements take place and are executed as planned, their generative systems can be created, maintained, and developed by independent programmers and smaller organisations as well.

Figure 2.3 shows the future trends in generative AI.

2.6 Evaluation and Quality Metrics

Evaluating generative AI is far more complex than assessing traditional machine learning models, since its outputs are often creative, open-ended, and context-dependent [18]. A combination of quantitative measures, qualitative judgments, and ethical audits is therefore necessary.

On the quantitative side, text models are evaluated with measures, such as perplexity, BLEU, and ROUGE, which capture fluency and similarity to reference

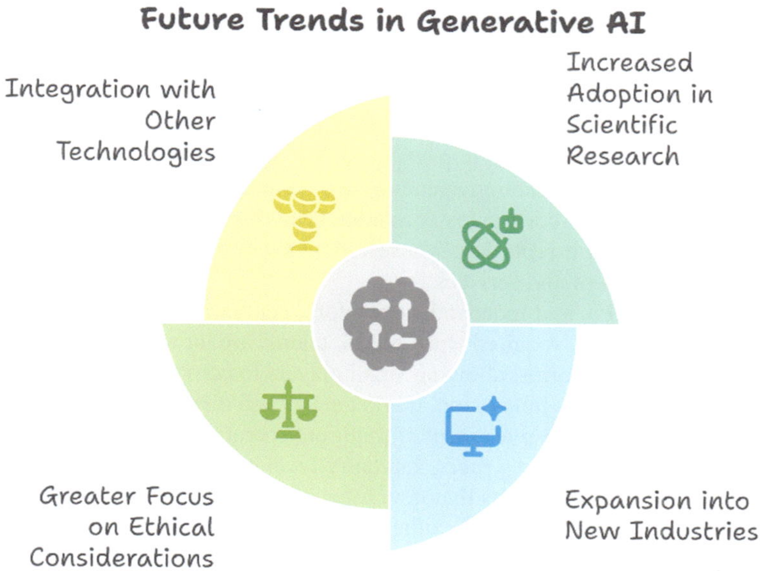

Fig. 2.3 Future trends in generative AI highlighting integration with other technologies, scientific research, ethical considerations, and expansion into new industries

outputs. For images, Fréchet Inception Distance (FID) and the Inception Score (IS) are commonly used to estimate visual quality and realism. These metrics are computationally efficient but often fail to capture subtleties such as originality, cultural appropriateness, or emotional impact.

To fill this gap, hybrid approaches combine automated scoring with human-in-the-loop evaluations. Human-in-the-loop evaluations involve expert humans in the process, performing crowdsourced user studies to assess the quality of generative AI content. For example, to understand the attractiveness of a new apparel, we cannot just rely on numerical evaluation metrics, but rather go through crowdsourced platforms and audit the human review.

Another growing area of such kind of evaluation is happening in the sector of ethical and bias auditing. This audit tries to identify stereotypes or exclusion in the patterns that lie in the training process of the generative AI.

Crucially, evaluation is not a one-time process. Because generative systems continuously evolve, dynamic and iterative evaluation frameworks are needed. Just as the models improve with feedback, the evaluation methodologies themselves must adapt to remain reliable and relevant.

2.7 Economic Impact Analysis

Generative AI is not only a technological breakthrough but also an economic disruptor, reshaping industries at both micro- and macro-levels [19].

At the organisational level, its impact is visible in reduced labour costs and shorter time-to-market cycles. Marketing agencies now generate dozens of ad variations in minutes, pharmaceutical companies accelerate drug discovery pipelines, and newsrooms automate routine reporting. These efficiencies lower costs while enabling faster scaling of content and services.

For startups, SMEs, and independent creators, generative AI is a democratising force. Tools that once required specialised teams are now accessible through subscription-based platforms, allowing small players to compete with resource-rich corporations. This fosters innovation, market entry, and diversity of products.

At the same time, the displacement of traditional creative roles cannot be ignored [20]. Graphic designers, copywriters, and entry-level programmers face automation risks. However, history suggests that new roles often emerge—such as AI trainers, prompt engineers, and human–AI collaboration specialists. Successful adaptation requires reskilling programs and proactive workforce policies that ensure workers transition into augmented rather than redundant roles.

On a macroeconomic scale, generative AI contributes to productivity growth and innovation cycles across advertising, media, entertainment, healthcare, and finance. Entirely new business models—such as personalised fashion-on-demand or AI-generated interactive media—are beginning to take shape.

Yet the economic landscape is complicated by evolving intellectual property dynamics. If AI generates a viral song or designs a marketable drug, who profits—the developer, the trainer, the user, or the AI's corporate owner? Emerging solutions include new licensing frameworks, attribution protocols, and revenue-sharing mechanisms, but global consensus is still lacking.

2.8 Summary

This chapter has mapped the journey of generative AI from its foundational architectures to its wide-ranging applications, ethical concerns, and economic implications. We examined the mechanics of GANs, VAEs, and newer paradigms such as diffusion and transformer-based models, highlighting how they have expanded AI's capacity for novel content creation across text, image, audio, and video.

From a practical perspective, we discussed the technical requirements and deployment challenges that shape successful implementations, along with case studies that demonstrated the value of generative AI in fields as diverse as filmmaking, fashion design, journalism, and gaming.

In this way, generative AI and general AI applications are actually helping human beings to create better art and other creative content.

This chapter summarises the requirements of responsible and ethical governance. Innovative AI is prone to be misused in the form of the generation of defects, propaganda, especially for political purposes, innovating hallucinated imperfect answers which appear as correct, conflicts in intellectual property and creating bias. Artificial intelligence generations and productions should be viewed through the lens of robust social, legal, and policy promoters.

References

1. Saxena D, Cao J. Generative adversarial networks (GANs) challenges, solutions, and future directions. ACM Comput Surv (CSUR). 2021;54(3):1–42.
2. Ahmad Z, Jaffri ZU, Chen M, Bao S. Understanding GANs: fundamentals, variants, training challenges, applications, and open problems. Multimed Tools Appl. 2025;84(12):10347–423.
3. Dash A, Ye J, Wang G. A review of generative adversarial networks (GANs) and its applications in a wide variety of disciplines: from medical to remote sensing. IEEE Access. 2023;22(12):18330–57.
4. Akkem Y, Biswas SK, Varanasi A. A comprehensive review of synthetic data generation in smart farming by using variational autoencoder and generative adversarial network. Eng Appl Artif Intell. 2024;1(131): 107881.
5. Chen H, Xiang Q, Hu J, Ye M, Yu C, Cheng H, Zhang L. Comprehensive exploration of diffusion models in image generation: a survey. Artif Intell Rev. 2025;58(4):99.
6. Thampanichwat C, Wongvorachan T, Sirisakdi L, Chunhajinda P, Bunyarittikit S, Wongmahasiri R. Mindful architecture from text-to-image AI perspectives: a case study of DALL-E, Midjourney, and stable diffusion. Buildings. 2025;15(6):972.
7. Hagos DH, Battle R, Rawat DB. Recent advances in generative AI and large language models: current status, challenges, and perspectives. IEEE Trans Artif Intell. 2024.
8. Huang D, Yan C, Li Q, Peng X. From large language models to large multimodal models: a literature review. Appl Sci. 2024;14(12):5068.
9. Zhang B, Wang T, Zhou C, Conci N, Liu H. Human trajectory forecasting using a flow-based generative model. Eng Appl Artif Intell. 2022;1(115): 105236.
10. Al-kfairy M. Strategic integration of generative AI in organizational settings: applications, challenges and adoption requirements. IEEE Eng Manag Rev. 2025.
11. Ferrara E. GenAI against humanity: Nefarious applications of generative artificial intelligence and large language models. J Comput Soc Sci. 2024;7(1):549–69.
12. Shoaib MR, Wang Z, Ahvanooey MT, Zhao J. Deepfakes, misinformation, and disinformation in the era of frontier AI, generative AI, and large AI models. In: 2023 International conference on computer and applications (ICCA); 2023. p. 1–7. IEEE.
13. Fontana AG. Intellectual property protection in the era of artificial intelligence and the problem of generative platforms. J World Intellect Prop. 2025.
14. Zhang Y, Gosline R. Human favoritism, not AI aversion: People's perceptions (and bias) toward generative AI, human experts, and human–GAI collaboration in persuasive content generation. Judgm Decis Mak. 2023;18: e41.
15. Jia Z, Chen J, Xu X, Kheir J, Hu J, Xiao H, Peng S, Hu XS, Chen D, Shi Y. The importance of resource awareness in artificial intelligence for healthcare. Nat Mach Intell. 2023;5(7):687–98.
16. Morante G, Viloria-Núñez C, Florez-Hamburger J, Capdevilla-Molinares H. Proposal of an ethical and social responsibility framework for sustainable value generation in AI. In: 2024 IEEE technology and engineering management society (TEMSCON LATAM);2024. p. 1–6. IEEE.
17. Sedkaoui S, Benaichouba R. Generative AI as a transformative force for innovation: a review of opportunities, applications and challenges. Eur J Innov Manag. 2024.

18. Bandi A, Adapa PV, Kuchi YE. The power of generative AI: a review of requirements, models, input–output formats, evaluation metrics, and challenges. Futur Internet. 2023;15(8):260.
19. Korinek A. Generative AI for economic research: use cases and implications for economists. J Econ Lit. 2023;61(4):1281–317.
20. Amankwah-Amoah J, Abdalla S, Mogaji E, Elbanna A, Dwivedi YK. The impending disruption of creative industries by generative AI: opportunities, challenges, and research agenda. Int J Inf Manag. 2024;1(79): 102759.

Chapter 3
Foundations of Quantum Computing: Concepts, Algorithms, and Real-World Challenges

3.1 Quantum Gates and Circuits

Similarly, these elements are quantum gates and circuits, as they are to classical computing gates and circuits. Superposition and entanglement introduce non-classical qualities; thus, the methods for designing and analysing quantum gates and circuits must be novel and entirely different from classical gate and circuit design and analysis. It is essential to deliberate on the possibilities of how the different comprehensive quantum gates may be combined to define quantum circuits in matters concerning future quantum calculations [1].

3.1.1 Quantum Gates

Like classical circuits, quantum gates play the same role as the classical building blocks of quantum Boolean circuits. Unlike them, however, qubits are designed very differently from classical computers, which handle Boolean logic operations. Classical gates can simultaneously represent 0 and 1 values in their superposition state to create a unified and highly generalised operation [2]. The properties of quantum gates are presented in Fig. 3.1.

- **The Pauli-X gate** is the quantum equivalent of the classical NOT Gate. It simply changes the state of the qubit. That means if the qubit is in the $|1>$ state, and vice versa. Because the Pauli-X gate allows basic state manipulation, it forms one of the most fundamental quantum computations.
- **Hadamard Gate**: This Hadamard table provides superposition. The Hadamard table accepts a qubit in either $|0>$ or $|1>$ and places both in the same superposition. That's why quantum computers can evaluate many calculations simultaneously, and how quantum parallelism works.

T. H. Sardar, *The Spectrum of Computer Science: Emerging Technologies and Trends*, Studies in Smart Technologies, https://doi.org/10.1007/978-3-032-19632-3_3

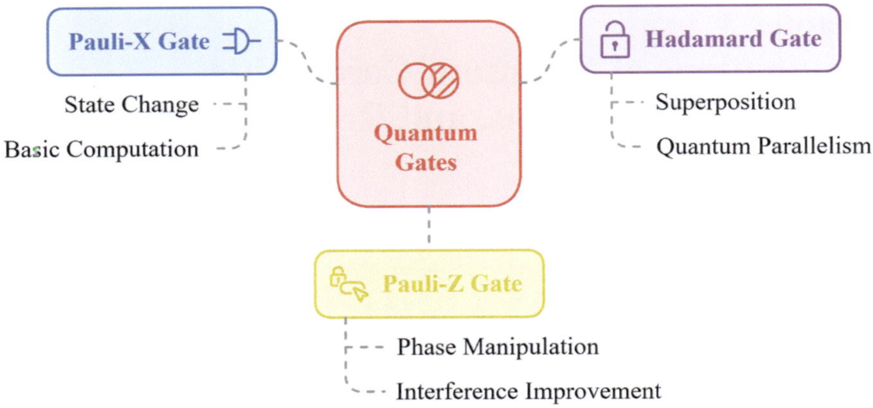

Fig. 3.1 Properties of quantum gates

- **Pauli-Z Gate**: A Pauli-Z gate changes the phase of a state in a qubit in quantum computers by pi radians. Although it modifies the phase of |0 and |1 states of the measured qubit, it does not alter the probability of the qubit being in any of the two states when a measurement is taken. Phase manipulation is crucial in many quantum algorithms where interference between quantum states improves the chances of correct responses while eliminating wrong ones. Sum states are used to enhance correct responses and eliminate incorrect responses.
- **Entangling Gates**: Entangling gates are among the crucial parts of quantum computing as they create entanglement between qubits—a phenomenon whereby the state of one qubit is intrinsically linked to the state of another. Entanglement is the principal cornerstone of quantum mechanics and underpins the power of quantum computing, allowing for phenomena such as superposition and quantum parallelism.
- **CNOT Gate**: The CNOT gate consists of two qubit gates that bind the qubits. It works as follows: When the first qubit (control qubit) is in one state, the gate switches on the second qubit, and the target qubit remains unchanged. Gates play an essential role in creating entangled states, where the state of one qubit is directly linked to another state regardless of the distance between them; this effect is essential in quantum computing and is used in quantum teleporters and hyper-dense coding gates for quantum circuits.
- **Toffoli and Fredkin Gates**: A Toffoli gate is an uncontrollable gate that changes the state of a target qubit only when both qubits are in the |1 state. A Fredkin gate changes the state of two fundamental modes depending on the state of the control qubit. These gates are essential for more advanced quantum algorithms and are part of the universal gate set for quantum computation.

An overview of quantum entanglement gates is shown in Fig. 3.2.

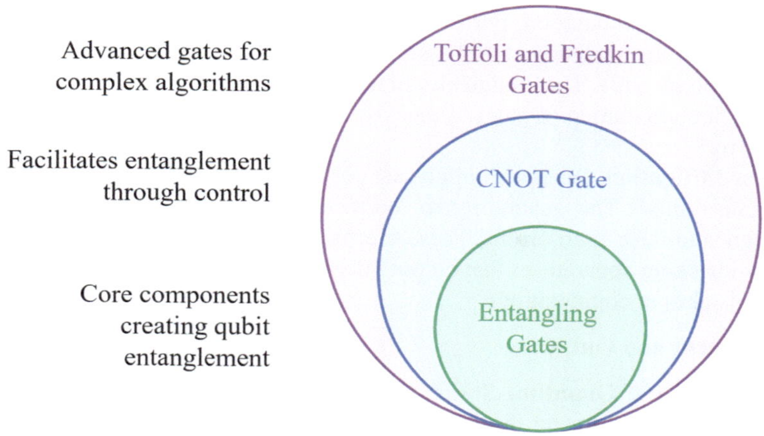

Fig. 3.2 Quantum entanglement gates

3.1.2 Quantum Circuits

A quantum circuit is a set of quantum gates that can perform specific calculations on qubits. The design of quantum circuits entails how the information will flow and interact in the quantum, and the potential for interference and decoherence [3]. The quantum circuit operation sequence is presented in Fig. 3.3.

Quantum Circuit Design

- **Initial Qubit**: A quantum cycle begins with one or more qubits, all initially set and most commonly to |0. This may be compared to declaring a variable and assigning it an initial value in classical programming.
- **Gate Application**: After initialisation, a quantum gate is used in a special section to control the qubits. The choice and order of the gates determine the outcome of

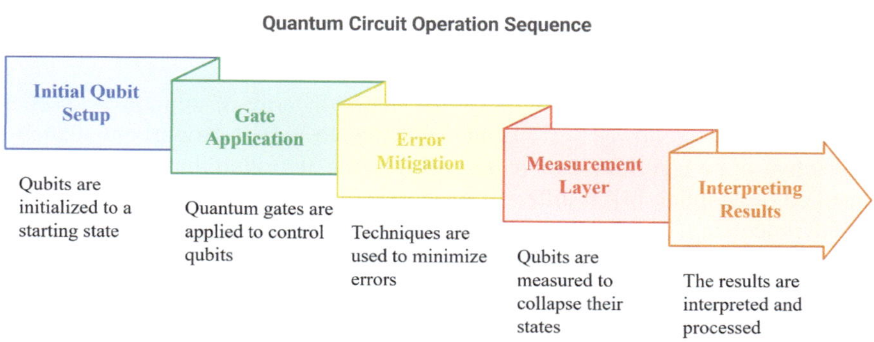

Fig. 3.3 Quantum circuit operation sequence

the quantum computation. For example, a circuit will first access a qubit with a Hadamard gate, then use a CNOT gate to access the qubit, and then use a Pauli-Z gate to phase shift. The complexity of a quantum circuit is created by choosing the right combination of gates to meet the required performance while reducing the error.
- **Error Mitigation**: Quantum circuits are perfect for decoherence errors and other quantum noise. The quantum error correction and the quantum error-tolerant design minimise such errors. These are more faulty qubits and gates, and the error-checking apparatuses that do not interfere with the quantum state to reduce the mistakes in computations.

Measurement and Output

- **Collapsing the Quantum State**: A measurement layer is the last layer of a quantum circuit. During the measurement of qubits, their previously superimposed quantum state yields a definite classical state, either (0) or (1). From these measurements, the value of quantum computation is determined. Since it is in different conditions simultaneously, a condition is being measured in terms of probability. This is also why, when one is using a quantum circuit where the same quantum circuit produces different results in different runs of the program, it is customary to run many algorithms to come up with a good result.
- **Interpreting Results**: A classical bit is a computation and is obtained from the measurement of quantum counterparts referred to as qubits. Hence, these results are only achievable based on the architecture of the quantum circuit deep within the quantum system. Sometimes, the final output of the quantum implementation of an inherently classical algorithm is post-processed by other classical algorithms, which are run on the outcome to extract information from it or to improve it in some way.
- **Image Suggestion**: An illustration of a quantum circuit involving qubits moving through gates, including Hadamard, CNOT, and Pauli-Z, with brief descriptions of each process.

3.2 Quantum Algorithms

As given earlier, quantum algorithms can be viewed in terms of solutions or a series of steps, which are followed by a quantum computer to address certain computations. A quantum of an algorithm can be described as the algorithm that is processed in a method of adoption [4]. It differs from classical algorithms at the quantum level of data processing. This section provided some knowledge on Shor's and Grover's algorithms and how quantum computers are helpful compared to classical computers.

3.2.1 *Shor's Algorithm*

Shor's algorithm is another quantum algorithm that has recently attracted much attention. It was developed by a mathematician called Peter Shor in 1994 [5]. Haber-Shoup optimally addresses one of the number one challenges in both mathematics and computer science: integer factorisation, which is the basis of many forms of cryptosystems such as RSA. The pros and cons of Shor's algorithm are provided in Table 3.1.

- **Factorisation and Cryptography**
  - **The Factorisation Problem**: A few issues are complex for regular computers, and one of them is decomposing a large integer into its corresponding prime factors. The former increases its difficulty as the size of the number grows exponentially, threatening it as a keystone of the contemporary. For instance, the RSA brilliant encryption method that is currently in use in almost all online communications is founded on the fact that, as much as one can quickly multiply two prominent prime numbers to obtain another equally prominent number, one will spend ages figuring out the prominent numbers back into the two original prime numbers using classical computers.
  - **Shor's Quantum Algorithm**: Quantum parallelism and interference help Shor's algorithm factoring large numbers to use a lot less time, in fact, tremendously. This is done by loading the period of a function associated with the number to be factored through the quantum Fourier transform instead of the exponential time that would take on the use of an ordinary computer, but which can otherwise be managed in polynomial time if a quantum computer is being used.
  - **Impact on Cryptography**: Although the capability of performing calculations in parallel and, therefore, much faster is a fact of today, Shor's algorithm could efficiently factor the standard public-key cryptography if a large-scale, very reliable quantum computer is available, which provokes a fundamental question. Because of this, post-quantum cryptography has emerged, focusing on developing a cryptographic system that may be safely deployed in both a quantum and a typical environment.

Potential Use Cases

Table 3.1 Pros and cons of Shor's algorithm

Pros	Cons
Fast factorization	Quantum computer dependency
Cryptanalysis potential	Threat to current cryptography
Prime testing	Implementation complexity
Quantum advantage	Limited practical use
Post-quantum cryptography	Resource-intensive

- **Cryptanalysis**: Shor's algorithm, apart from the above findings, can be applied in any other cryptographic system that employs the factorisation problem, hence shifting the security of data once more.
- **Prime Number Testing**: Shor's algorithm is also applicable in number theory as it tests for the primality of large numbers, which is valuable in various scientific and "crypto"-related uses.

3.2.2 *Grover's Algorithm*

It provides the quantum search method with a quadratic improvement over classical methods, thus making it a significant breakthrough in quantum computing. Grover's algorithm is a quantum algorithm that is designed primarily to find unstructured search problems, or to put it another way, search for unsorted databases [6]. Figure 3.4 provides Grover's algorithm search process.

Concept and Background

- **Classical Search Problem**
  - **Unsorted Database Search**: In the classical computer, it was observed that if the database is not sorted, then the number of operations that are required to search for a particular target item among N items has to be significant, as each item requires attention. In the worst-case instance, this takes about O(N) time, where N is the size of the database of items. Besides, in large databases, this may take quite a lot of time and may not even be efficient since one must

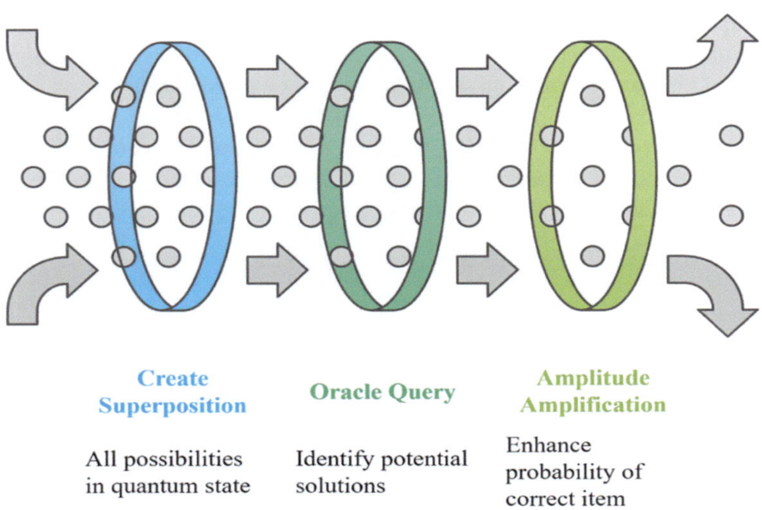

Fig. 3.4 Grover's algorithm search process

search for each item until the target one is found or all the items that have been queried.

- **Unstructured Search**: An unstructured search is when a person looks for an item or a solution to a problem in a dataset that does not appear to have any structure at all. Unlike network databases, where indexes or sorts assist in searching operations or acceleration, unstructured data requires a scan of all the candidates.

- **Grover's Quantum Algorithm**
  - **Quantum Superposition**: Grover's algorithm uses other forms of quantum features, such as entanglement and interference, for instance, in the use of amplitude amplification to perform a search of a database, in contrast to classical computer algorithms. Instead of checking each item individually, Grover's algorithm creates a superposition of all possible items and applies quantum operations to amplify the probability of finding the correct item.
  - **Amplitude Amplification**: An unstructured search is when a person looks for an item or a solution to a problem in a dataset that does not appear to have any structure at all. Unlike network databases, the likelihood of getting the wrong answer is reduced. This is done through a series of quantum operations that afford a step-by-step improvement of a particular piece of the quantum state.

Algorithm Steps

- **Initialisation**
  - Creating superposition: We first create the quantum state of all possibilities from GROVER's algorithm. This is made through the use of a quantum register where each of the qubits is in the state 0 and 1 at the same time. Each of the qubits goes through a Hadamard gate, which places all the NNN states at a probability amplitude of 1/2.

- **Oracle Query**
  - **Oracle Function**: An oracle is a quantum algorithm that determines the presence of the solution to the speech problem. It is supposed to negate your state of amplitude in the target state while being otherwise indifferent to other states. This is done by phase flipping the right solution in the quantum state space, with a tag, to enable identification of the correct solution.
  - **Black-Box Operation**: The oracle merely scans the input and writes into the input also as a read–write black box that checks or compares against the target item. For each state, it says if it is applied to the target state and thereby decides if the phase has to be inverted—information that is crucial for the next step.

- **Amplitude Amplification**
  - **Inversion About the Mean**: This is explicitly done by inversion about the mean in Grover's algorithm to increase the amplitudes of the state corresponding to the oracle query. This step takes the difference of each step, which equals its

amplitude divided by its average amplitude and takes the minus for the value obtained. The objective is to enhance the probability of quantifying the state and to achieve this by amplifying the target state concerning other states.
- **Iterative Process**: In Gover's algorithm, the oracle query and the amplitude amplification are iterated several times. The number of iterations would be $\sqrt{n}$ where N stands for the number of achievable states. It is about several attempts to facilitate the possibility of achieving a measure of the right solution.

- **Measurement**
  - **Extracting the Result**: After following the iterations, an attempt is made to take measurements on the quantum register. Therefore, the target state is likely to be observed because of the amplitude amplification. The solution to the search problem is given in the measurement process.

Performance and Applications

- **Quadratic Speedup**
  - **Efficiency Compared to Classical Methods**: The algorithms presented by Grover are fast and provide a quadratic advantage over the classical search methods. While a classical search is said to take O(N) operations, Grover's algorithm runs through O(N) operations. This quadratic enhancement was exhibited when operating on big datasets; this guarantees the existence of quantum algorithms.

- **Applications**
  - **Database Search**: Grover's algorithm can be applied for any general unstructured search, for example, searching for records in the databases for records, patterns, or items. This has potential implications for data retrieval, information extraction, and other database-related tasks.
  - **Optimisation Problems**: Grover's algorithm can also be applied to all those tasks that require finding the best solution among all possible options. For example, it can be used for carrier logistics, financial budgets, and scheduling series where the best search option is relevant.
  - **Cryptography**: Using the brute force search method, the algorithm can be applied to cryptographic systems. Grover's algorithm may have the chance to reduce the time used in cracking specific forms of encryption, as it may improve the algorithms of the exhaustive search, such as controls and considerations.
  - **Quantum Hardware Requirements**: Such an algorithm can be amplified only with the help of quantum hardware that uses many qubits and accurate quantum operations. While Grover's is efficient for the present quantum computers, these computers pose problems in the number of qubits and coherence time.

3.2.3 *Real-World Quantum Computers and Limitations*

The domain of quantum computing is actively progressing in leaps and bounds, with several companies and study organisations pushing the limits. This phase of the work presents the present-day state of matters regarding technological quantum hardware, achievements so far registered, and the mountains that remain to be climbed [7].

3.2.3.1 Current State of Quantum Hardware

Types of Quantum Computers [8]

(a) **Superconducting Qubits**: Superconducting qubits are implemented using materials that become superconductors around absolute zero. Such materials allow for the coherent manipulation of quantum states, enabling the implementation of quantum logic gates and circuits, as presented in Fig. 3.5. Also, examples are provided below:

- **IBM's Q System One**: IBM quantum computer employs superconducting qubits and seeks to provide a robust environment to test out quantum codes and apps.
- **Google's Sycamore Processor**: In 2019, Google's Sycamore processor defeated all other supercomputers worldwide in solving a particular task, demonstrating quantum supremacy for the first time. This device is also built using superconducting qubits and is designed to perform complex computations.

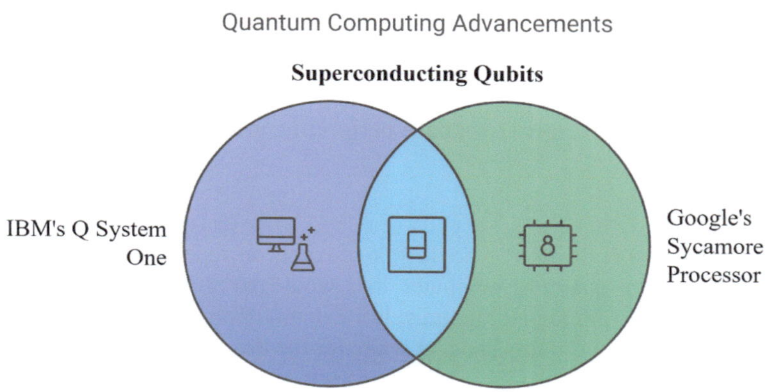

Fig. 3.5 Quantum computing advancements—superconducting qubits

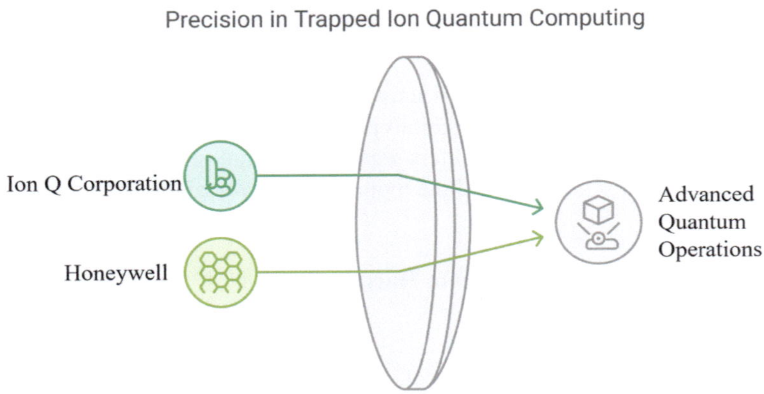

Fig. 3.6 Precision in trapped-ion quantum computing

(b) **Trapped-Ion Qubits**: Trapped-ion qubits involve the use of magnetically confined ions, which are then manipulated with lasers to perform quantum operations. This method is known for its accuracy and stability. Figure 3.6 provides the precision in trapped-ion quantum computing.

- **Examples**
  - **Ion Q**: This Corporation specialises in trapped-ion quantum computers, which might as well be known because of their high reliability and scaling potential. Ion's designs are tactically laser-assisted to control qubits with great precision.
  - **Honeywell**: Similarly, Honeywell's trapped-ion quantum computer systems are also excellent and are known for their precision as well, being deployed on the genesis of novel quantum algorithms and applications.

(c) **Topological Qubits**: Topological qubits are built on a theoretical fragment, an anion with some exceptional hypothetical quantum attributes. This technology provides further strength in performing error correction and tolerance to decoherence.

- **Examples**
  - **Microsoft Station:** Microsoft's research on topological qubits aims to design a quantum computer that would be more resilient and fault-tolerant. Although it's tentatively about this approach, however, the further development of this concept waits for the quantum computer hardware.

3.2.3.2 Milestones in Quantum Computing

(a) **Quantum Supremacy**: The year 2019 was also significant in improving quantum computing because it was the year that Google claimed quantum supremacy, the capacity of the machines to accomplish specific operations that were possible only by the fastest classical supercomputers. This may be the first time attempts were made to show that quantum computing had capabilities other than what could be achieved by classical computing. Quantum communication and the ability to allow practical quantum computers can now be discussed because of this milestone, which emphasises the need for the emergence of more quantum computation in general.

(b) **Early Quantum Algorithms**: The efficient and highly coordinated nature of small-scale quantum processors was used to incorporate and model quantum algorithms on diverse fronts. Applying quantum simulations and optimisations includes chemical processes that have displayed great performance in problems related to drugs and materials. The focus of the catalytic machines also brings out aspects of any quantum computers in solving real-world problems, though the current size of quantum gadgets limits this.

3.2.3.3 Challenges and Limitations of Quantum Computing

(a) **Quantum Decoherence**: Quantum decoherence occurs because quantum information is highly vulnerable to noise in its environment, which dispels the necessary quantum superpositions and entanglements for computation [9]. Sustaining Q coherence against decoherence is a principal tenet in quantum tech. Turning now to addressing decoherence, it is at the top of the list when staying as quantum computations attempted to remain faithful to what was calculated needs protection from decoherence, which interferes with the necessary quantum states having to preserve themselves so as to last as long as the computation. But if you do not correct decoherence, qubits become decoherent, and worse, the computation dissipates.

(b) **Error Correction**: Other quantum error correction codes include the so-called surface codes and cat codes, where errors are corrected slowly, without a direct probe to the quantum state, which collapses it. These methods incorporate additional qubits and circuitry onto the platform.

3.2.4 Practical Applications and Industry Impacts

(a) **Materials Design**: Quantum computing is the material science proper to a new fundamental matter. In other words, fancy new materials for various electronics and cutting. Quantum computing is thus the material science appropriate to a new fundamental matter. Or, in other words, pretty new metals for that equipment to oppose various electronics or advanced technology such as renewable energy. That would allow searching for new materials or degrading material in a more conducive way (or similar), making new tech all across aerospace to consumer electronics. Subsectors include renewable power generation as an edge tech. That would allow for finding new materials or deleting the material in a more conducive way (or similar), which would open up new technologies everywhere, from aerospace to consumer electronics.

(b) **Environmental Modelling**: Abstract quantum computing, as we've seen with its ability to simulate complex chemical systems, is a natural fit for the computational problems in environmental science. This includes identifying and forecasting pollutants, ecological variables, weather patterns, etc. Better models will allow us to address and solve global issues like climate change and pollution.

(c) **Optimisation Problems** [10]

- **Logistics**: Quantum computing will be primarily applied to solve logistical issues like route finding, supply chain, and inventory, and it will yield good results because many resources will be invested in such matters. On the one hand, productivity levels would rise, and, consequently, services would become both faster and cheaper if there were quantum computers that would enhance the delivery-made routing. They would also improve the supply chain by minimising wastage by stocking the merchandise according to the expected demand, which is also on merit.
- **Finance**: Although recent technologies are fundamentally about risk management, which may be magnetic and successful, sundry optimisations that are pretty standard in this area are demonstrated more by quantum algorithms. With these large datasets, these companies' quantum investments could be appropriately deployed on departments or business obligations in the best way. Besides this, it will also aid in fraud detection by involving extraordinary computations that the current computer system cannot process.
- **Transportation**: Some of these are called transport problems, and computation can then be applied to solve combinatorial issues, such as traffic light control or car routing. Real-time control of traffic lights, which involves the application of a quantum algorithm, can be deployed. Hence, proportionate reduction of congested vehicle traffic and general traffic enhancement will be easily possible. On the one hand, productivity levels would increase, and therefore services would be faster and cheaper when quantum computers were invented, improving the routing of deliveries. They would further enhance the supply chain by avoiding wastage by efficiently stocking the merchandise based on a demand forecast.

3.3 Summary

This chapter provides a comprehensive introduction to quantum computing, detailing its core principles, foundational technologies, and the practical challenges it faces. It begins by distinguishing quantum computing from classical computing, highlighting the use of quantum bits (qubits) in superposition and entanglement to perform calculations. The text then delves into the building blocks of quantum computation: quantum gates and circuits. It explains the functions of key gates, like the Pauli-X, Hadamard, and CNOT gates, and describes how these are combined into circuits to perform complex calculations while addressing the unique challenges of error correction and decoherence.

The chapter then shifts its focus to two of the most significant quantum algorithms: Shor's algorithm and Grover's algorithm. It explains how Shor's algorithm poses a threat to modern cryptography by efficiently factoring large numbers, and how Grover's algorithm provides a quadratic speedup for unstructured search problems. The final sections provide a realistic look at the current state of quantum hardware, including superconducting qubits and trapped ions, and discuss the major milestones and ongoing limitations, such as quantum decoherence and the need for robust error correction. The chapter concludes by exploring the vast potential of quantum computing in real-world applications, from materials design and environmental modelling to solving complex optimisation problems in logistics and finance.

References

1. Van de Wetering J. Constructing quantum circuits with global gates. New J Phys. 2021;23(4): 043015.
2. Lin D, Xiang Z, Xu R, Zhang S, Zeng X. Optimized quantum implementation of AES. Quantum Inf Process. 2023;22(9):352.
3. Levenson-Falk E, Shanto SA. A review of design concerns in superconducting quantum circuits. Mater Quantum Technol;2025
4. Jayakumar A, Adedoyin AA, Ambrosiano JJ, Anisimov PM, Bärtschi A, Casper WR, Chennupati G, Coffrin CJ, Djidjev HN, Gunter DO, Karra S. Quantum algorithm implementations for beginners. ACM Trans Quantum Comput. 2022;3(4):18.
5. Opiłka F, Niemiec M, Gagliardi M, Kourtis MA. Performance analysis of post-quantum cryptography algorithms for digital signature. Appl Sci. 2024;14(12):4994.
6. AbuGhanem M. Characterizing Grover search algorithm on large-scale superconducting quantum computers. Sci Rep. 2025;15(1):1281.
7. Desdentado E, Calero C, Moraga M, García F. Quantum computing software solutions, technologies, evaluation and limitations: a systematic mapping study. Computing. 2025;107(5):1–60.
8. Proctor T, Young K, Baczewski AD, Blume-Kohout R. Benchmarking quantum computers. Nat Rev Phys. 2025;7(2):105–18.
9. Rishiwal V, Agarwal U, Yadav M, Tanwar S, Garg D, Guizani M. A new alliance of machine learning and quantum computing: concepts, attacks, and challenges in IoT networks. IEEE Internet Things J. 2025.
10. Deleplanque S, El Yaagoubi A, Gras T, Descamps F, Mourrier J, Lefebvre I. Applying analog quantum computing to solve optimisation problems: case studies on Max-Cut and CVRP. Int J Syst Sci: Oper Logist. 2025;12(1):2463530.

Chapter 4
Data Analytics and Machine Learning Insights

4.1 Data Preprocessing and Feature Engineering

Machine learning is a branch of artificial intelligence. Machine learning techniques are trained on a dataset such that they can provide a systematic solution without being explicitly guided every time. For example, if you develop a suitable machine learning model and train it with the pictures of cats and dogs. Then the model can distinguish the features of cats and dogs and build a systematic solution to later identify cats and dogs from the new image datasets that contain either a cat or a dog or both.

However, model training is very sensitive to the quality of underlying data [1]. The real-world datasets usually contain noise and errors. Data preprocessing techniques are used to remove this noise and errors from the data, and to reduce the size and transform its structure, based on the requirements and nature of the dataset. Future engineering, on the other hand, creates new and useful features (data variables) from that clean data [2]. Processing and feature engineering steps are very important for a machine learning model to work effectively.

The data preprocessing steps are shown in Fig. 4.1 using a funnel structure.

4.1.1 Data Preprocessing

Data preparation is a systematic process which makes the raw data ready for machine learning algorithms. It removes inaccuracies, incompleteness, and noise and standardises the data values for the machine learning algorithms to work efficiently [3].

Data cleaning is one of the important data preparation techniques where we remove uncertainty and wrongly entered data values from the dataset and make it ready for the machine learning algorithms. Raw data quality is sometimes compromised due to many factors. Sometimes the factors are due to human error in entering

T. H. Sardar, *The Spectrum of Computer Science: Emerging Technologies and Trends*, Studies in Smart Technologies, https://doi.org/10.1007/978-3-032-19632-3_4

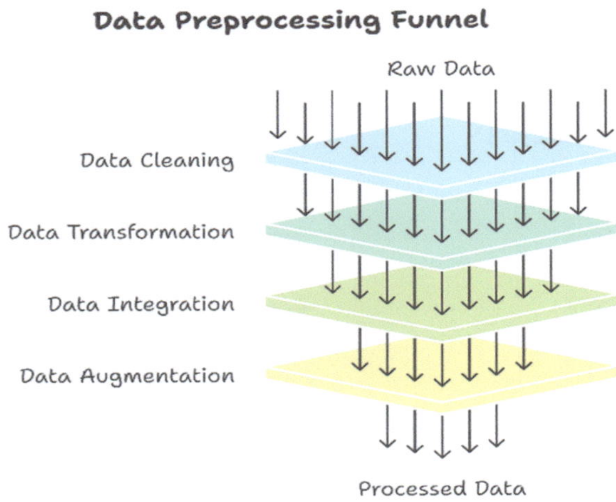

Fig. 4.1 Data preprocessing funnel from raw data to processed data through key stages

the data, or the transformation of the database during transmission, or sometimes due to the IoT device or sensor malfunction. These mistakes create missing data values in a dataset, sometimes make aeronauts' entries in the dataset or make duplicate entries, which are very prone to generating unsatisfactory results if directly provided to the ML algorithms.

To solve the issue of missing values, we can use manual input, such as specifying the missing values. There, we can provide a certain number, for example, 999, to those places where the values are missing. Using statistics, we sometimes use the mean and median to solve missing numerical values. At the same time, we can use the Mode value for categorical variables. However, in recent times, the generative AI models such as variational encoders and generative adversarial networks have proven to be very useful for solving the missing value problem.

Removing duplicates is a challenging task for datasets where the values and their inter-differences are narrow. For such kinds of datasets, manual and traditional statistical methods are proven not to be so effective. In this scenario, we can use probabilistic machine learning approaches, such as support vector machine and random forest, to use a similarity matrix to identify duplicate values and remove them efficiently.

The data transformation technique of data preprocessing is used to ensure compatibility among different variables of a particular dataset. Sometimes, machine learning techniques, for example, clustering, face a significant challenge if two variables in a particular dataset carry a different range of values.

One of the most important data transformation techniques is normalisation. Sometimes the data distribution is very large, and we have to make the data distribution in a certain range for the machine learning algorithms to work effectively. This important job is done by normalisation. Scaling is a data transformation task that normalises the range of a particular data variable into a fixed range from 0 to 1. Similarly, the

Z-score normalisation tries to fix around 0 with unit variance. This particular normalisation is very important for those algorithms that assume a normal distribution of dataset values. This flexibility is not easily available for a categorical variable. We can create discrete ranges for the normalisation of the categorical variable instead.

In today's world, an Enterprise can consist of multiple servers and other sources of data generation. This data is sometimes required to be gathered together for a machine learning task. This can be achieved by data integration. In data integration, we gather data from multiple sources while preserving the property of the data, while maintaining the schema of the data source, and to make sure that the semantics and the data values' units are not changing.

Data augmentation is a technique where we add features or samples to the data, or we add more noise to the data, to finally increase the size and diversity of a dataset. This makes data modelling easier and prevents overfitting for a particular machine learning algorithm. In data reduction techniques, we reduce the size of the dataset while preserving the important features and statistical qualities of the data, so that the computational cost of executing the dataset, as well as the time, can be saved.

4.1.2 Feature Engineering

Future engineering is a process to convert the raw data attributes into a meaningful and interpretable set of attributes to enhance the machine learning models' performance and computational efficiency. This can only be done with the help of expertise in that dataset, relevance of the attributes for a data mining job.

Figure 4.2 shows how raw data is refined using different stages, including the feature engineering step.

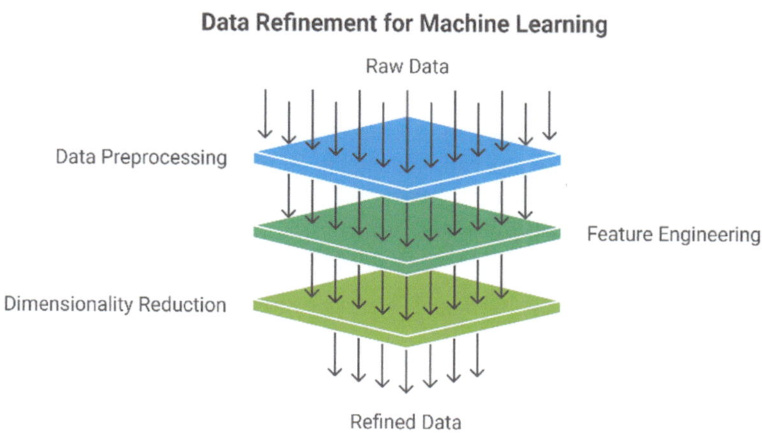

Fig. 4.2 Data refinement process through preprocessing, feature engineering, and dimensionality reduction

Feature creation is one of the important techniques of feature engineering. Sometimes in the raw datasets, there are relationships and patterns among the variables. At the creation level, we understand those relationships and, based on that, create new variables or features and add them to the data. This will enhance the machine learning model's performance. New features sometimes capture new insights in the dataset that individual features don't reveal. For example, a dataset of house prices can have multiple attributes such as square feet and number of rooms, but may not have a feature named as living square feet. This feature can be newly created and added to the dataset as a multiplying resultant of the previously named features.

Feature subset selection is a data dimensionality reduction technique where we select only the minimal required features from a large set of features available in a dataset. Feature selection is very useful as it reduces computational time, improves model performance, and saves the time of developers to develop a particular machine learning task. For this purpose, statistical techniques, such as correlation analysis, mutual information, and information gain, can be used. Also, wrapper methods and filter methods are very popular methods in feature subset selection.

Dimensional reduction is a popular feature extraction method where the high-dimensional data is transformed into a lower-dimensional space. This is performed while retaining key information of the dataset. This helps in saving storage space and processing time. There are two most popular methods for dimensionality reduction: principal component analysis and wavelet transform. t-Distributed Stochastic Neighbour Embedding (t-SNE) provides non-linear dimensionality reduction particularly effective for data visualisation, preserving local neighbourhood structures to reveal clustering patterns, and non-linear relationships.

4.2 Machine Learning Algorithms

The machine learning algorithms are the backbone of a machine learning ecosystem, where useful and relevant patterns, insights and descriptions of a dataset, and predictive analysis come out of the input data. There is a large set of algorithms which can be used for different purposes of machine learning objectives and dataset types. Choosing machine learning algorithms depends on their suitability for a particular dataset, a certain application, and ranges from implementing linear statistical methods to highly complex deep learning techniques.

Machine learning algorithms can be broadly categorised into three generic groups: supervised, unsupervised, and reinforcement learning [4]. This concept is shown in Fig. 4.3.

Fig. 4.3 Key categories of machine learning algorithms: supervised, unsupervised, and reinforcement learning

4.2.1 Supervised Learning: Classification

Supervised learning is one of the popular tasks of machine learning. In a supervised machine learning model, the model is trained on a labelled dataset. A labelled dataset is a kind of dataset where, along with the feature values of n input variables, the output/target label values are also available. The job of the model is to understand for which particular ranges of values of the input variables, the output or the target variable's value is determined. Upon successful model training (known as "learning"), a supervised machine learning model can accurately predict the target variable values for new instances of the input variables. In most classification models, the outcome is binary: the classification target can be divided into two classes. In more complex classification models, the classification can also get into multi-class: the output is classified into more than two classes. However, in the case of multilevel classification, each data tuple instance for input variables may belong to more than one output class. An example of classification is spam detection, where an incoming email is classified into two classes: spam and not spam. The machine learning supervised algorithm is trained on a label dataset where each email is tagged as either spam or not spam based on input variables such as the sender information, the specific keywords in the content of the email, the behavioural pattern of the sending IP address, and the metadata. If the incoming email has the matching value corresponding to this input variable, then the model will recommend the particular email as spam.

Figure 4.4 provides a graphical view of the classification algorithm and its appropriate use.

Linear classification techniques, such as logistic regression and linear discriminant analysis, rely on the assumption that different classes can be separated with a straight line or hyperplane. One advantage of these methods is that they are easy to interpret, helping us see which features influence the predictions most. Logistic regression estimates the probability that a data point belongs to a given class. Support Vector Machines (SVMs) take a different approach by identifying the optimal boundary that maximises the margin between classes, providing strong performance even with

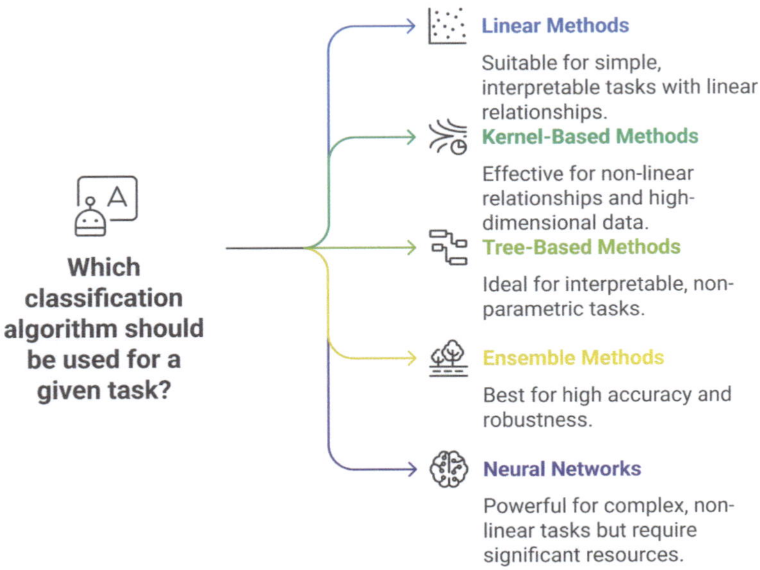

Fig. 4.4 Best suitable classification algorithm

small datasets. Using kernels, SVMs can also handle non-linear data patterns by projecting inputs into higher-dimensional spaces, with popular kernels including polynomial, radial basis function (RBF), and sigmoid. Decision trees break data down into a series of simple rules, making them very transparent, although deep or complex trees may overfit. To overcome this, ensemble approaches like random forests and gradient boosting combine multiple models for better accuracy and stability. Random forests construct numerous trees using random samples and subsets of features, which reduces overfitting while allowing insight into feature importance. Gradient boosting builds models sequentially, each one correcting the errors of the previous, often yielding state-of-the-art results in structured datasets. Neural networks, as highly flexible function approximators, excel at modelling complex, non-linear relationships. Deep learning architectures, such as convolutional neural networks (CNNs) for images and recurrent neural networks (RNNs) for sequential data, have greatly advanced performance across many fields, although they remain challenging to interpret in sensitive applications.

4.2.2 Supervised Learning: Regression

Regression is another popular supervised learning technique. Regression is designed to work with numerical data. Unlike the classification, supervised learning methods are designed for categorical data. Regression algorithms are used to predict continuous numerical outcome values. This is particularly useful for application areas such as performance optimisation in industries, demand forecasting for products of an e-commerce portal, and price prediction.

There are many regression models available. In regression, we try to obtain the linear correlation in terms of statistical features between input variables and target variables. In regularised regression, we address overfitting in the dataset. On the other hand, the non-linear regression models try to retrieve complex relationships which is not possible to retrieve by linear regression models. Popular examples of non-linear regression models are radial boosting, random forest, and decision trees.

4.2.3 Unsupervised Learning: Clustering

The last two subsections discuss supervised learning, where the target or outcome variables are already available in the dataset, and the target variable prediction is easier for the models, as it's based on the "already available" input data value-based training. However, in many real-world scenarios, the class/output levelled feature is not available in the dataset. In such cases, can we use predictive analysis using supervised learning? No. Rather, we can go for descriptive analysis using unsupervised learning techniques. For example, consider you have been provided the unlabelled data of different kinds of sweets. Sweets are rasgulla, jam, mysore pak, and bundi. How will you differentiate between all these sweets as they are not labelled? In such cases, we have to use an unsupervised learning mechanism where we learn from the data features regarding the similarities between the same data items and gather them together in a single place.

Clustering is the most popular and supervised learning technique. Clustering is a supervised machine learning technique where the similarities between the data objects are calculated based on similarity measures, and then the similar data items are gathered together in one place, known as a cluster. In this way, we can create n number of clusters from the input dataset. Clustering has a wide application in solving real-world problems. For example, applications in anomaly detection, detecting network intruders, organising data based on their properties, and enabling exploratory analysis.

The most popular casting algorithm is the K-means clustering algorithm. This is a partition-based clustering algorithm. In this algorithm, first, we determine how many clusters to create for a particular input dataset. Then, that many data points are randomly obtained from that dataset, and then start iteratively calculating similarities with all other datasets. Most similar data points are attached to the initially selected data points based on the highest similarity to create a number of clusters.

Hierarchical clustering creates nested cluster structures through agglomerative or divisive approaches, providing dendrograms that offer insights into data structure at multiple scales. Density-based clustering methods, exemplified by DBSCAN, identify clusters as regions of high data density separated by low-density regions, excelling at discovering irregularly shaped clusters while automatically determining cluster numbers.

Gaussian mixture models assume data originates from mixtures of Gaussian distributions, providing probabilistic cluster assignments rather than hard partitions. Spectral clustering transforms data into lower-dimensional spaces using eigenvectors of similarity matrices, enabling identification of non-convex clusters particularly effective for image segmentation and social network analysis.

4.2.4 Reinforcement Learning

Reinforcement learning addresses sequential decision-making problems where agents learn optimal behaviours through interaction with dynamic environments. Unlike supervised learning, RL algorithms learn from rewards and penalties rather than explicit input–output pairs, making them particularly suitable for control systems, game playing, robotics, and autonomous systems.

The reinforcement learning framework models interactions as Markov decision processes characterised by states, actions, transition probabilities, and reward functions. Value-based methods, including Q-learning and Deep Q-Networks (DQNs), estimate state-action pair values to guide decision-making, with neural networks approximating Q-functions for high-dimensional state spaces.

Policy-based methods directly optimise policies mapping states to actions, proving effective for continuous action spaces. Advanced techniques like proximal policy optimization incorporate constraints to ensure stable policy updates and prevent performance degradation during training.

4.3 Applications of Data Analytics and ML

Practical applications across diverse industrial sectors demonstrate the transformative potential of data analytics and machine learning for operational efficiency, decision-making enhancement, and innovation acceleration [5–10]. As illustrated in Fig. 4.5, data analytics applications evolve along a spectrum from reactive tasks such as regulatory compliance and fraud detection to proactive strategies including predictive maintenance and drug discovery.

4.3.1 Healthcare

Healthcare applications address critical challenges in diagnosis, treatment optimization, drug discovery, and health system management. Medical imaging represents one of the most successful applications, with convolutional neural networks achieving

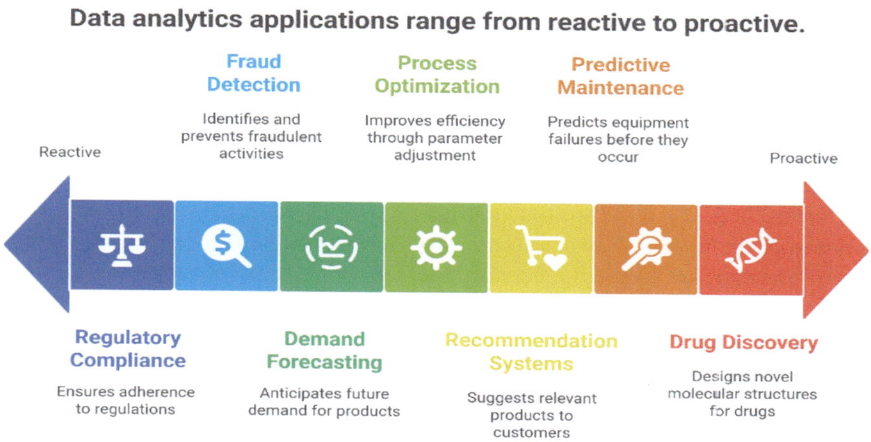

Fig. 4.5 Data analytics applications spanning from reactive approaches to proactive strategies

expert-level performance in radiology, pathology, and dermatology for conditions ranging from diabetic retinopathy to skin cancer detection.

Electronic health records provide rich longitudinal data for predictive modelling and clinical decision support. The natural language processing techniques can automatically retrieve knowledge from clinical notes, doctors' prescriptions, and lab reports and reactively identify a patient's condition or adverse drug reactions, etc.

In the same way, machine learning algorithms can be used for analysing genomic structures in large datasets and can be used for prescribing Precision medicine. The generative AI models can be used for describing beneficial modules from a set of chemicals and can also suggest a suitable chemical composition for new drug Discovery. Generative AI can also be used for identifying the toxic elements in a particular drug, and can significantly reduce the side effects by removing those elements.

4.3.2 Finance

The machine learning-based data analytics are proving to be very useful for financial applications. Financial applications are prone to fraud, and they require excessive automation for customer service and for managing the risk in terms of financial losses.

The risk management for loans and investments is now using machine learning-based data analytics to determine which users are potential defaulters. This can be achieved by supervised learning on existing financial variables of a person's or a company's credit history-based financial datasets.

Fraud protection systems also use supervised machine learning algorithms to learn from historical frauds and their associated variable value ranges to predict the future potential of a fraud. Machine learning algorithms can also generate graph-based data structures, which can help enhance the visualisation and assessment of existing coordinator frauds.

Share market pricing and predictions are also using supervised machine learning algorithms for determining profits and losses, and to provide users with a better understanding of the market. For day-to-day stock trading and long-term stock trading, these algorithms are providing increasing reliability to the users and financial firms. Dialogue's best systems are benefiting from automatic financial monitoring systems.

4.3.3 Retail

The retail firms use algorithms for their automated inventory management, understanding the pricing of their product, the working style of their organisation, and to enhance the customer experience for better engagement and retention of customers. The retail firms are also focused on demand forecasting. In demand forecasting technique, the system tries to understand the trends of the market, the preferences of the users and seasonal factors that can promote the sales and advertising of their product. Machine learning algorithms can learn from historical data and can provide support and automation in all these tasks.

The modern retail operations also rely on recommendation systems. The recommendation systems are designed to learn user behaviours, previous preferences, their online browsing pattern on the website, and purchase history so that future predictions on their personalised product preferences can be easily made. Similarly, price optimisation is also provided by a machine learning algorithm which automatically adjusts product prices, whether in a single item or in bulk sales, based on the other market competitors and their price listing, and market demand based on season and festivals.

The unsupervised machine learning technique of clustering is also used in the retail segment as the clustering algorithms can segregate the users and customers based on their online behaviour and preferences, their region, and other facilities they are preparing from a certain retailer. Targeted marketing becomes easier for the retail firms with the segregation of users, as their dynamics become easier to understand.

4.3.4 Manufacturing

Industry 4.0 mandates the manufacturing companies to use automated software for maintenance of the manufacturing items, quality control, optimising the automated processes, and supply chain management.

The machine learning algorithms, trained on IoT sensor data, can provide predictive maintenance for manufacturing units, materials, and equipment. A successfully designed predictive maintenance model can predict and proactively manage equipment failures and can prescribe different working parameters. For example, it can suggest suitable temperature arrays, vibration ranges, and suggest digital twin technologies.

Machine learning, along with computer vision techniques, is used for the detection of quality control-related defects with a higher range of accuracy without involving human intervention. This significantly reduces the cost of maintenance and enhances quality outcomes. Similarly, process optimisation algorithms that use the reinforcement learning technique of machine learning can automatically learn from their previous failures and identify the best suitable parameters for operating a state of manufacturing instruments in real time.

4.3.5 Transportation

The transportation sector uses digital systems for traffic management, enhancement of user safety, employment of autonomous bodies, and providing the root of optimisation.

Machine learning algorithms can learn from historical data and predict proper traffic management. It can provide a better understanding of a particular set of roads which are dependent on transport and can provide dynamic scheduling of traffic lights. It can also provide the best suitable shortest path route for individual vehicles. Also, it can provide a larger lending route, which can save time and fuel efficiency by avoiding traffic.

Machine learning, with the computer vision technology, can make its own decision on driving control using multiple sensors built into vehicles and process it in real time. The machine learning technique of reinforcement learning is also used in this context.

4.3.6 Education

The machine learning algorithms also help the education sector by providing a personalised learning experience. It can also help improve the learning outcome by providing students with analytics of their learning outcome using different graphs and other visuals, by processing and learning from student interaction and performance data. Machine learning-based tutoring systems are also being deployed and tested, which can deliver learning content and teaching materials based on individual student capacity and learning behaviour.

4.3.7 Energy/Smart Grids

The energy sectors are heavily focusing on integrating traditional energy and renewable energy sources into the smart grid. It is also required for grid optimisation and forecasting the demand of energy at a specific time of day, a week, or in a particular season. The smart grid technology requires proper two-way interactions between the energy producers, the energy distributing companies, and the energy consumers, so that the best comprehensive understanding of demand can be predicted.

The renewable energy sector poses a huge challenge as it depends on the weather and many other external factors. The modern machine learning algorithms can learn from historical and weather forecast data and provide those learning outcomes to the renewable energy sector. Similarly, optimisation algorithms are used to provide optimal supply and demand of energy so that the grid can maintain stability.

4.4 Evaluation Metrics for Machine Learning Models

The reliability of a machine learning model cannot be proven until it is evaluated using a suitable evaluation metric [11–13]. Based on the type of machine learning technique used, and the dataset or the application and its objective, a suitable evaluation method is required to determine its effectiveness. Sometimes, more than one evaluation metric is used for evaluating a single or a set of machine learning models for a particular task, and based on the majority voting of their opinions, we finally choose a particular machine learning model.

For classification-based machine learning models, metrics such as accuracy, recall, and F1-score play a vital role in determining the effectiveness and applicability of a machine learning model.

The accuracy metric justifies the correctly predicted instances out of the total predictions made by that model. For example, a spam classification model which predicts 80% of the spam from the inbox while the actual number of spam emails is just 40%, then the accuracy of the spam classification model is 50%.

Precision, on the other hand, justifies the correct number of truly predicted instances out of all the predictions. Similarly, provides an evaluation of how many true positive cases are actually identified. For example, in the prediction of a cancer classification system, high recall is required as it is required to be sure that all cancer cases are properly predicted, although some false negatives occur. F1-score is the harmonic mean of precision and recall. It balances both precision and recall. For example, in a customer review system, F1-score is very crucial as it can provide the proper balance of proper classification of positive and negative sentiments.

In regression evaluation, mean absolute error and mean squared error play a crucial role. Mean absolute error provides the mean value of the absolute difference between the actual values and the predicted values. For example, if we find the mean absolute error value is 5000 euros in a house price prediction system, the model predictions are off by the same amount on average. Mean squared error squares the error before calculating the mean value. It is designed in this way to penalise the larger deviation severely than the deviation with a smaller margin. For example, predictive values which are deviated far from their actual value should be highly penalised in an energy consumption prediction system so that energy mismanagement and wastage can be prevented.

For clustering evaluation, usually, the ground truth is not available. Clustering algorithm's evaluation happens best on the internal matrix. Some of the popular internal matrices are Silhouette, Dunn, and Davies–Bouldin, etc.

For clustering evaluation, usually, the ground truth is not available. Plastering algorithm's evaluation happens best on the internal matrix. Some of the popular internal matrix are still how to Davies–Bouldin, etc. A matrix usually tries to assess the cohesion and separation inside a cluster and between the clusters, respectively.

4.5 Challenges in Data Analytics and Machine Learning

There are many challenges faced by machine learning and data analytics techniques. These challenges can be broadly classified into three parts: technical, management, and ethical challenges. Data quality challenges, a technical challenge, include the issue with training, machine learning algorithms with raw data, which contains noise, biases, inconsistencies, different semantics for the same features, and missing data values. A machine learning algorithm which runs on this high-quality data is prone to giving incorrect results. Scalability issues are another major technical challenge that machine learning algorithms face. Modern data has shifted to the big data sphere, where the volume, variety, and velocity of the data challenge machine learning algorithms in making correct predictions and summarisations. Management challenges mainly deal with the unavailability of expert professionals in this domain, which can delay projects and provide poor quality results. The ethical challenges are mainly based on the wrong uses of machine learning algorithms, which can affect humanity in general in a negative way [14–17].

4.6 Advances and Emerging Trends

The field of machine learning is experiencing a quick growth. In this, Automated Machine Learning (AutoML) is a technique which can choose machine learning algorithms automatically, select their suitable parameters, and adjust with a proper model based on the ensemble-based majority voting of the evaluation metrics. In this way, automatic machine learning reduces the requirement for direct expert intervention, enhancing the development of more models in a shorter time.

Federated learning is another domain which is growing rapidly and developing at a very fast rate. Transfer learning-based machine learning techniques can learn

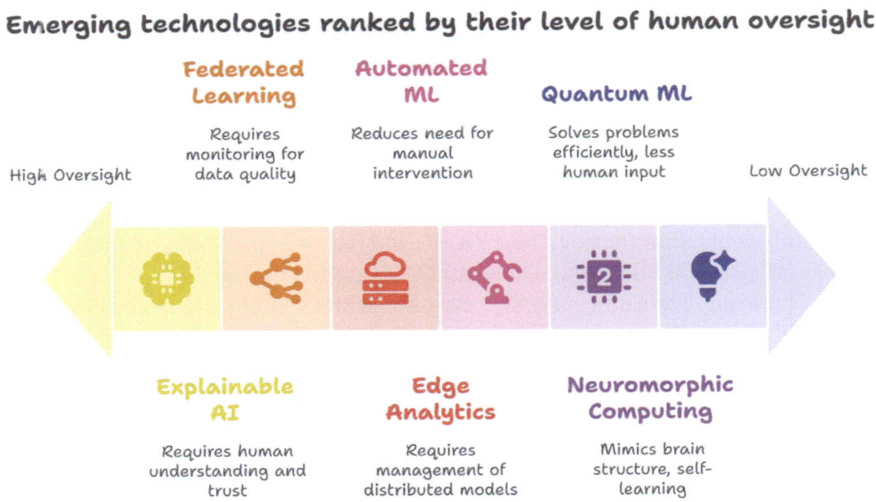

Fig. 4.6 Emerging technologies ranked by their level of human oversight

from training datasets while the data sources are spread across different sites. This process executes while maintaining the privacy of the users and their data. Federated learning is applied on domains such as finance, healthcare, and mobile devices.

In the same line, we have explainable AI, which is a technique to explain a machine learning model. It ensures detailed explanations for an outcome to occur by a machine learning model for a particular set of inputs. Thus, explainable AI develops trust in the machine learning model. Quantum machine learning explains the use of quantum computing in the machine learning set-up for computational speedup. Techniques such as Transformer architectures are being used for making efficient natural language processing, and generative artificial intelligence is being used for generating images, videos, and text at a very fast but reliable speed. Self-supervised learning algorithms are reducing the need for labelled data for a model to predict future labelling of a data input.

Figure 4.6, emerging technologies are ranked by their level of human oversight.

4.7 Ethical, Legal, and Social Implications

Legal frameworks like GDPR in Europe and CCPA in the U.S. protect data by ensuring privacy, transparency, and user consent, while industry-specific rules such as HIPAA, SOX, and FCRA safeguard healthcare, financial, and credit data, respectively. Ethical concerns include algorithmic bias, privacy violations, lack of transparency, and job displacement due to automation. Socially, technology offers benefits like improved healthcare and personalised education but also raises challenges

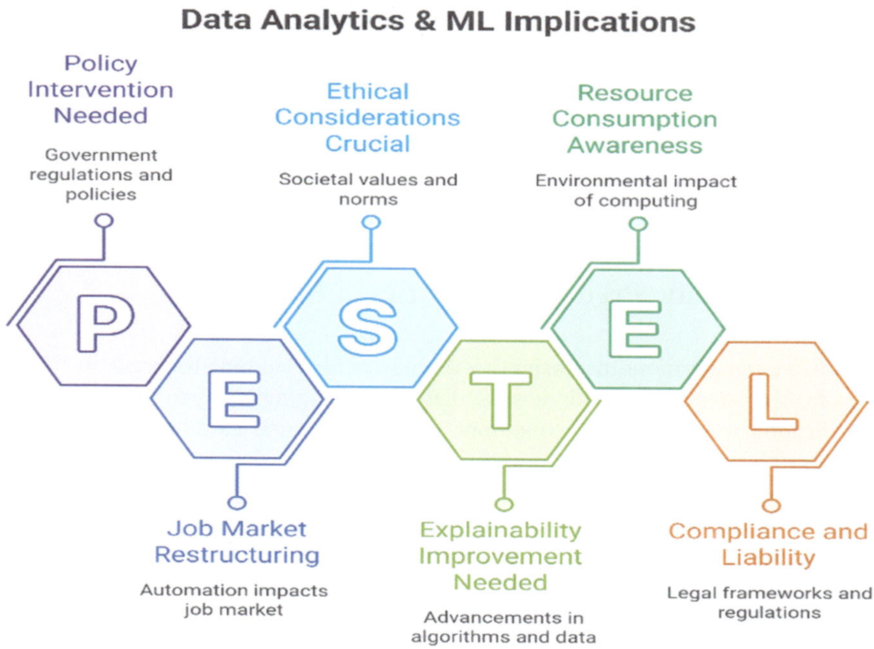

Fig. 4.7 PESTEL framework highlighting key implications of data analytics and machine learning

such as the digital divide, echo chambers, and mass surveillance affecting individual freedom.

Figure 4.7, the PESTEL framework captures the broader implications of data analytics and machine learning.

4.8 Tools and Frameworks for Data Analytics

Data analytics and machine learning can be implemented by many programming languages. However, the Python programming language has to have a special mention in this context. The Python programming language consists of packages, such as Pandas, Scikit-learn, and NumPy, which are very suitable for developing machine learning algorithms and applications. Similarly, TensorFlow and PyTorch are useful for deep learning-based application development. R Studio is a tool specially designed for statistical analysis and visualisation, and is widely used by machine learning developers. SQL can also be used for querying from the machine learning-based analysis of structured datasets.

Many cloud platforms are used for machine learning development. AWS Sage-Maker has to be given a special mention for its speciality in end-to-end machine learning pipelines, Google Cloud Vertex AI is popular for unified machine learning platforms, and Microsoft Azure Machine learning Studio is used for visual model designing and deployment at a large scale [18].

Big data processing utilises Apache Hadoop for distributed storage and processing, Apache Spark for in-memory computation acceleration, and Apache Kafka for real-time data streaming in event-driven architectures.

4.9 Case Study: Predictive Analytics in Healthcare

Healthcare organisations increasingly leverage machine learning for predictive diagnostics using electronic health records, diagnostic imaging, and genomic data. Risk stratification models predict conditions like diabetes and cardiovascular disease before clinical symptoms appear, utilising logistic regression, random forests, and gradient boosting for accurate forecasting [19].

Practical applications in healthcare demonstrate how machine learning and data analytics can directly enhance patient care. Hospital systems, for example, predict patient readmission rates by combining structured data such as lab results and medication histories with unstructured information from physician notes and imaging reports. In addition, real-time monitoring systems powered by platforms like Apache Kafka allow continuous tracking of patient vitals in intensive care units, automatically generating alerts for any signs of critical events. By integrating advanced analytics, scalable processing frameworks like Apache Spark, and comprehensive visualisation dashboards, healthcare providers create holistic ecosystems that not only improve the quality of patient care but also achieve measurable reductions in healthcare costs through preventive interventions and more efficient resource allocation [20].

4.10 Summary

This chapter explored the world of data analytics and machine learning, showing how these technologies can extract meaningful insights and create value from a wide variety of datasets. It started with the basics, like data preprocessing and feature engineering, and then moved into the core algorithms of supervised, unsupervised, and reinforcement learning, which power today's predictive and descriptive analytics.

Real-world applications were highlighted across sectors, such as healthcare, finance, retail, manufacturing, transportation, education, and energy, illustrating how data-driven approaches can reshape decision-making and drive innovation. The chapter also discussed evaluation metrics to ensure models perform effectively, while addressing practical challenges like handling big data, ensuring scalability, interpreting complex models, and managing ethical concerns.

Recent developments, including AutoML, federated learning, deep learning advances, and MLOps, were examined, along with their ethical, legal, and social implications. Practical tools and frameworks were shared to guide implementation, and a healthcare case study demonstrated tangible benefits, from improving patient care to optimising resources.

By blending theory with practical examples, this chapter provides a complete view of modern data analytics and machine learning, highlighting their transformative potential while stressing the importance of responsible and ethical deployment.

References

1. Das S, Sardar TH, Sahana DS. An ensemble technique using genetic algorithm and deep learning for the prediction of rice diseases. In: Machine learning hybridization and optimization for intelligent applications: CRC Press;2025. pp. 289–303.
2. Sudarshana K, Yendapalli V, Kamala L, Sardar TH, Adhoni ZA. CADFRA: Coronary artery disease feature reduction with autoencoder for optimistic and effective classification. In: Owoc ML, Varghese Sicily FE, Rajaram K, Balasundaram P, editors. Computational intelligence in data science. ICCIDS 2024. IFIP advances in information and communication technology, vol. 717. Springer, Cham;2024. https://doi.org/10.1007/978-3-031-69982-5_15.
3. Sardar TH, Khatun A, Sengupta S, Alam Y, Ara T. Machine learning in the healthcare sector and the biomedical big data: techniques, applications, and challenges. In: Big data computing;2024. pp. 336–52.
4. Subbarao E, Gite P, Sharma A, Vyakaranam L, Sardar TH, Taqui SN, Ganeshan P, Metwally AS, Kalam MA. Investigation of solar energy for urban areas and metropolitan communities using machine learning. Electr Power Compon Syst. 2023; 1–11.
5. Sardar TH, Ansari ZA, Theerthagiri P, Karthikeyan P, Ayyasamy V, Saini DKJB. MapReduce-enhanced fuzzy K-Least medians for qualitative clustering of document big data. Concurr Comput: Pract Exp. 2025;37(4–5): e70035.
6. Swasthika Jain TJ, Sardar TH, Sammeda Jain TJ, Guru Prasad MS, Naresh P. Facial expression analysis for efficient disease classification in sheep using a 3NM-CTA and LIFA-based framework. IETE J Res. 2025; 1–15. https://doi.org/10.1080/03772063.2025.2498610.
7. Theerthagiri P, Ruby AU, Chandran JGC, Sardar TH, Shafeeq BMA. Deep SqueezeNet learning model for diagnosis and prediction of maize leaf diseases. J Big Data.2024; 11(1):112.
8. Sardar TH. Reflecting on a decade of evolution: MapReduce-based advances in partitioning-based, hierarchical-based, and density-based clustering (2013–2023). Wiley Interdiscip Rev: Data Min Knowl Discov.2024; e1566.
9. Bellary MZ, Deepthi THS, Musthafa BA, Ahmed SJ, Sarkar R. Medical image analysis of knee Osteoarthritis using modified deep CNN. J Surv Fish Sci. 2023;10(2S):133–44.
10. Sardar TH, Hazarika RA, Pandey B, Prasad G, Hassan SM, Dodmane R, Gohel H. Video key concept extraction using convolution neural network. In: 2024 IEEE 3rd international conference on AI in cybersecurity (ICAIC). IEEE;2024 pp. 1–6.
11. Sardar TH, Sarkar R, Ahmed SJ, Bandyopadhyay A. A novel ensemble methodology to validate fuzzy clusters of big data. In: Proceedings of the fourth international conference on trends in computational and cognitive engineering: TCCE 2022. Singapore: Springer Nature Singapore;2023. pp. 267–78
12. Sardar TH, Pandey BK, editors. Big Data Computing: Advances in Technologies, Methodologies, and Applications. CRC Press;2024. https://doi.org/10.1201/9781032634050.
13. Bellary MZ, Deepthi TH, Sardar B, Aziz Musthafa SJ, Ahmed SR. Medical image analysis of knee osteoarthritis using modified deep CNN. J Surv Fish Sci. 2023;10(2S):133–44.
14. Hassan MAM, Sardar TH, Fahim MFH, Mohamed MS, Suleyman RM, Usman MM. Artificial intelligence and deep learning applications on big data computing frameworks. In Big data computing. CRC Press;2024. pp. 196–211.

15. Khatun A, Sardar TH. Cryptography algorithms to prevent different security attacks. In Next generation mechanisms for data encryption. CRC Press;2025. pp. 35–46.
16. Gupta S, Singh U, Rawat M, Kurian A, Mandal L, Gupta P, Sardar TH. Best image processing for higher face detection rate using Haar cascades. In: International conference on robotics and networks. Cham: Springer Nature Switzerland;2023. pp. 113–28.
17. Kumar C, Kumar KA, Gupta S, Sardar TH. Cyberbullying detection based on the fusion of DistilBERT and SIMHASH technique. In: 2024 2nd international conference on artificial intelligence and machine learning applications theme: healthcare and internet of things (AIMLA). IEEE;2024. pp. 1–4
18. Chandarana P, Vijayalakshmi M. Big data analytics frameworks. In: 2014 international conference on circuits, systems, communication and information technology applications (CSCITA). IEEE;2014. pp. 430–4.
19. Khalifa M, Zabani I. Utilizing health analytics in improving the performance of healthcare services: a case study on a tertiary care hospital. J Infect Public Health. 2016;9(6):757–65.
20. Mishra HV, Kumar S. Predictive analytics: a case study in healthcare for lifestyle diseases. Math Eng Sci Aerosp (MESA). 2021; 12(4).

Chapter 5
Artificial Intelligence and Ethics

5.1 Introduction

Artificial intelligence and its importance are well known. It has a great impact on modern industries, financial systems, and society. AI systems are solving industrial automation and quality control issues, providing solutions for detecting financial fraud and forecasting profitability, automating traditional routine tasks and enabling complex decision-making. However, with all these benefits, AI also bring many minors to severe ethical concerns and a few unintended negative impacts after deployment [1].

The ethics in AI deal with the appropriateness in terms of transparency of the AI model, generation of output in a bias-free manner, accountability, and the human environment in the autonomous process. Ethical measures and their requirements have become imperative due to the role of modern AI systems in making critical decisions in sensitive sectors like justice and healthcare [2].

The objective of this chapter is to provide an overview of ethical issues with the development of modern AI and its applications. This chapter will provide multiple ethical challenges of the AI applications in different domains. It also discusses the requirement of suitable ethical frameworks to guide an AI application which complies with the ethical standards.

5.2 Foundations of Artificial Intelligence

To understand the ethical challenges associated with AI applications, one must clearly understand the technical foundations and theorems of artificial intelligence, which form the foundation of ethical issues. In this section, we provide different types of artificial intelligence systems, their core components and lifecycle considerations [3].

T. H. Sardar, *The Spectrum of Computer Science: Emerging Technologies and Trends*, Studies in Smart Technologies, https://doi.org/10.1007/978-3-032-19632-3_5

5.2.1 Types of AI

Artificial intelligence is a variety of different systems that can affect differently and work in solving problems in different ways. Categorise these different types of artificial intelligence systems into three broad categories:

Narrow AI or Weak AI: The weak AI systems are AI algorithms and models designed to do a certain task only. For example, the artificial intelligence systems that can only perform natural language processing, or an AI system which can only navigate an autonomous vehicle to a destination, are considered as a weak AI system. They are considered weak as they can only perform within a predefined domain and within its limited domain. Weak AI systems are not designed for general reasoning capabilities. The ethical considerations for the weak AI systems lie in the misuse of the application and its domain-specific biases.

General AI or Strong AI: The general AI applications are considered to have critical thinking, problem-solving, automatic learning for a provided new environment, and general reasoning capabilities. These self-learning capabilities make general AI perform intellectual tasks with a human-like capacity. General AI, or strong AI, is a hypothetical (largely theoretical) AI which is not yet been implemented in the real world. The ethical considerations for the strong AI systems lie in the potential societal disruption when the general intelligent devices are integrated into human societies, the immediate loss of many existing jobs, the erroneous behaviour and fraudulent attitude that these AI products can make and become unidentifiable by human beings.

Super AI or Superintelligent AI: A super AI or superintelligent AI is a futuristic concept where we consider that AI-based applications and models will be so advanced that they can surpass human intelligence in general. The superintelligent AI will have the capacity to think beyond the capacity of a human; it can solve problems which humans cannot solve, and it can provide critical thinking and out-of-the-box discoveries that humans cannot perform to date. Intelligent AI systems present significant ethical considerations. Beyond the control over the human mind, to go beyond the human accountability, it possesses the ultimate ethical concern: the existential risk for the human race.

5.2.2 Core Components of AI

To mimic human-like intelligence, a machine should be enabled with proper interoperability of many artificial intelligence building blocks. These technically driven building blocks are complex in nature, and ethical oversight occurs frequently [4].

The backbone of modern AI systems lies in the successful deployment of machine learning models. Machine learning models learn from data automatically and provide summarisation and prediction as well [5] (Fig 5.1). Key categories include the following:

- *Supervised Learning*: Supervised learning is one of the popular tasks of machine learning. In a supervised machine learning model, the model is trained on a labelled dataset. A labelled dataset is a kind of dataset where, along with the feature values of n input variables, the output/target label values are also available. The job of the model is to understand for which particular ranges of values of the input variables, the output or the target variable's value is determined.
- *Unsupervised Learning*: Unlike supervised learning, an unsupervised machine learning technique learns from the unlabelled datasets. The data objects are gathered together in a single place based on the features that they share.
- *Reinforcement Learning*: Reinforcement learning addresses sequential decision-making problems where agents learn optimal behaviours through interaction with dynamic environments. Unlike supervised learning, RL algorithms learn from rewards and penalties rather than explicit input–output pairs, making them particularly suitable for control systems, game playing, robotics, and autonomous systems.

Natural Language Processing is a technique of machine learning which enables machines to read, send, and generate natural language of human beings. Techniques of natural language processing are automated translation from one human language to another, analysing the sentiments from the human texts, conversation using chatbot agents, etc. In the domain of natural language processing, ethical issues are mainly concentrated towards misinformation generation, political propaganda generation,

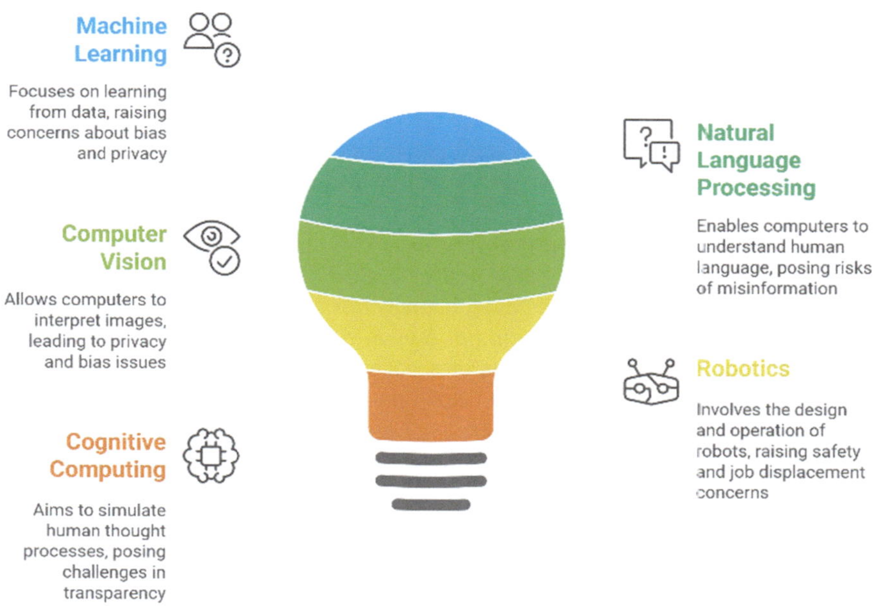

Fig. 5.1 A diagram showing the main components of AI (ML, NLP, computer vision, robotics, cognitive computing) with associated ethical concerns

spreading hate speech, chatbot-based fraud, the generation of automatic mysterious reviews, etc. [6].

Computer vision is a machine learning technique where images and videos are analysed for obtaining real knowledge and decisions. The applications of computer vision are focused towards surveillance systems, medical imaging, and automated vehicles. Ethical issues involve privacy concerns, recognition system biases, and potential civil liberties infringements [7].

Robotics is a machine learning technique that combines intelligent algorithms with physical machines to perform real-world tasks. Its applications range from industrial automation and warehouse management to healthcare assistance and disaster recovery operations. However, robotics also raises important ethical concerns, such as a huge part of traditional jobs are becoming replaceable due to automation, and ensuring safety during human and robot collaboration. Another important risk of robotics stems from decision-making in sensitive environments like defence or emergency response [8].

Cognitive computing refers to the use of computer systems that simulate human thought processes. Cognitive computing makes computers analyse complex problems and experience-based learning for swift decision-making. The applications of cognitive computing include financial advising, personalised learning, and medical decision-support tools. Ethical concerns in this domain consist of the availability of human biases within algorithms, the reliability of decision-making in sensitive domains, and the replacement of intellectual workers [9].

5.2.3 AI Lifecycle and Ethical Implications

Developing an AI system requires multiple components with their complex interoperator requirements. These complex components are required to be implemented with ethical consideration [10]:

The data collection and pre-processing phase required consideration of the ethical implications, such as consent-based retrieval of the dataset, ensuring the user's privacy, and equity. When the data is obtained from a sensitive part of the human population, it can result in a violation of human rights if the proper consent is not taken. It is also be taken care that the biases should be removed from the datasets. If these ethical considerations are not taken into account while collecting the data, it would result in inaccuracies and biases.

During the *machine learning model development*, it is required to train the underlying datasets. This training can result in the generation of region-based, religion-based, and other ethnical biases. These biases would create a significant social and political impact. Training large machine learning models can negatively impact environmental sustainability as it requires huge energy consumption.

During the phase of *deployment and monitoring* of the model, the real-world societal effect becomes prominently seen. We have to monitor and ensure that the model is not drifting from its real purpose and causing biases and manipulations. It must include the users and other stakeholders of the AI application to enable them

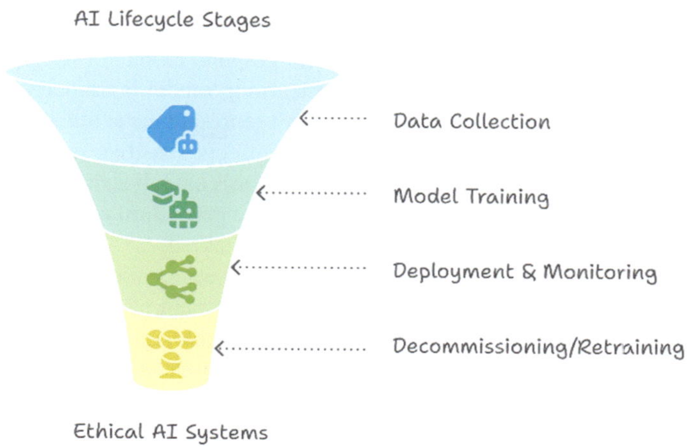

Fig. 5.2 Sequential stages of the AI lifecycle

with a feedback mechanism so that they can report any societal negative impact immediately.

The machine learning models *retraining and retiring* should always be kept activated to accommodate new international rules and regulations, and update the government laws and societal changes. If the AI models continue to provide obsolete algorithms, then it will attract societal harm and involve governments and enterprises in banning the AI application or mandating lifecycle management practices.

Figure 5.2 illustrates the AI lifecycle stage. At the top, it starts with data collection and goes down using steps such as model training, deployment and monitoring, to retraining at the bottom. In all these stages, ethical implications should be noticed and corrected accordingly.

5.3 Ethical Challenges in AI

AI application in different domains of work carries ethical implications [11].

5.3.1 Bias in AI

The bias in AI can be seen in different forms, such as historical bias, measurement bias, and sampling bias.

Historical bias occurs in an AI application when the AI model training is performed on historical data, as these datasets consist of past societal inequalities. For example, a decision-making artificial intelligence system can provide bias against credit defaulters for a particular region where credit defaulters were seen decades ago. The same thing can be seen against a demographic, ethnic, or religious group.

Sampling bias occurs when the machine learning model is trained on a dataset that consists of a population sampling that doesn't represent the actual characteristics of the population. Sampling bias inadvertently affects sensitive domains such as criminal justice systems and healthcare systems. In this case, if a small group doesn't get the representation that it deserves to receive, the prediction or judgment can be skewed to one side, and the existing biases can be retained.

Measurement bias occurs when the machine learning model is trained on a particular standard which prefers only a particular segment of the population while ignoring the broad diversity of the population. For example, if an AI system is trained on Caucasian facial recognition image datasets, a machine learning application will not provide accurate classification of facial identities of other ethnicities.

5.3.2 *Fairness in AI*

Fairness in AI requires not only technical improvements and accuracy but also ethical considerations towards creating systems that are inclusive and non-discriminatory.

Fairness audits are carried out to carefully check how AI systems perform across different groups of people and contexts. By regularly examining outcomes and identifying patterns of unequal treatment, these audits help uncover bias and provide guidance for corrective actions.

Fairness-aware algorithms are being actively developed to directly reduce bias during the training process. This can be done by modifying the loss functions, using regularisation techniques, or adding fairness constraints into the optimisation process.

Embedding fairness principles means making fairness a core part of AI development from the beginning. This includes encouraging inclusivity in organisational processes, ensuring that diverse perspectives are represented, and setting clear institutional benchmarks that define what fairness should look like in practice.

5.3.3 *Transparency in AI*

Transparency in AI ensures that the individuals of the application trust the system, as they would know how the application and its components work. Transparency is difficult to achieve due to the black-box nature of the AI applications [12].

Explainable AI (XAI) provides methods to make the decision-making process of AI more transparent and understandable to people. Explainable AI systems are more important for sensitive domains such as Healthcare justice. Explain the ability of every error to build up more distrust in the applications.

Proper documentation through tools also plays a key role. These documents provide a detailed explanation of the performance measures in the training data

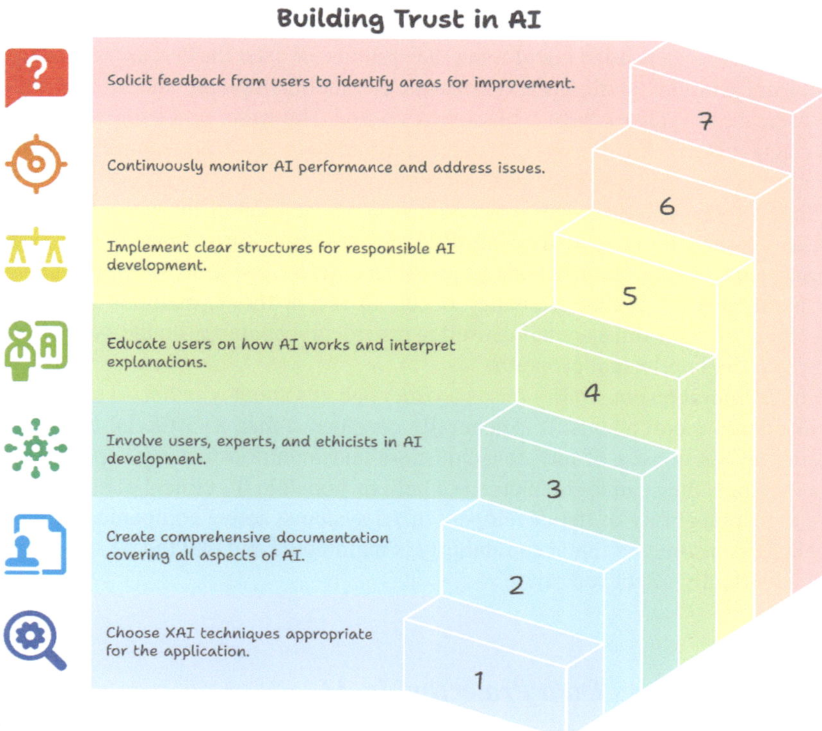

Fig. 5.3 Stepwise approach to establishing trust in AI systems

and their limitations. This would build trust in the model training and ultimately in the public who uses that AI application.

Finally, user *interrogability* provides a chance for the users to ask questions or to pose a query for validating the output. Interrogability enhances trust in an AI application as it provides accountability even when the machine learning models are performing a complex task.

Figure 5.3 shows a step-by-step bottom-up pathway for building trust in AI systems.

5.3.4 Accountability and Responsibility in AI

Assigning accountability for AI systems is one of the most difficult ethical issues because the autonomy and complexity of these systems make it unclear who should be held responsible when things go wrong [13].

From a legal perspective, responsibility is generally placed on human actors such as developers, organisations deploying the systems, or even the end users. The traditional legal artificial intelligence systems are now being implemented for the masses. However, these systems have a black-box nature. The users don't understand how these legal models actually work. It makes it difficult to prove and identify casualties when the system wrongly provides an output.

From a technical perspective, accountability is supported through mechanisms such as detailed logging of training data, decision-making pathways, and system environments. These records make it possible to trace how a decision was reached and to investigate failures after they occur. In recent times, methods like impact assessment of algorithms are being used to provide transparency on the system while protecting the intellectual property.

Finally, moral responsibility raises broader debates about whether AI systems can be considered moral agents. It is generally considered that, as artificial intelligence systems do not consist of any consciousness and intention-making capacities like human beings, they cannot be blamed or held responsible for ethical considerations. It is the responsibility of the developers of AI systems, active companies, and policymakers to ensure that fair accountability is maintained and that the AI system does not inadvertently affect society.

5.3.5 Privacy and Data Protection in AI

AI relies heavily on large amounts of data, which naturally raises serious concerns about privacy and data protection. The way data is collected, stored, and reused determines not only technical efficiency but also ethical responsibility [14]:

Data Minimisation and Consent: A basic principle in this area is collecting only the minimum amount of data that is truly necessary, and doing so with explicit, informed consent from users. In practice, however, AI often works with datasets that were collected in different contexts, making it difficult to guarantee proper consent or compliance with laws such as GDPR. Ethical data governance, therefore, requires clear communication about how data will be used, along with honest disclosure of potential risks.

Deidentification and Reidentification Risks: Protecting personal information is very important. For this purpose, anonymisation and pseudonymisation techniques are usually used. But even these techniques cannot always protect personal information. Using modern artificial intelligence tools, which are designed to link multiple datasets, can achieve maximum accuracy in the reidentification of a person.

Surveillance and Social Control: There are many artificial intelligence-enabled surveillance tools used by gated communities, law and order enforcement organisations, and government agents to track and monitor a large segment of citizens or populations. Although many of these uses are ethical and for the safety needs of the people, they always carry ethical implications. For example, misuse of these surveillance tools may threaten the freedom of expression, spy on the masses, and discriminate.

5.4 Responsible AI Development

To address the issues discussed in the last section, we need responsible development of artificial intelligence systems which follow the ethical guidelines using user input and feedback, expert opinions and government rules and regulations for the development of the tools [15].

5.4.1 Ethical Guidelines and Principles

There is a list of ethical guidelines and principles developed by policymakers and AI researchers. Principles are aimed at making AI tools and applications should not (1) negatively affect the users, (2) can benefit in gaining social good, (3) providing an automated decision-making system to the experts and the users of a particular system, and, most importantly, (4) to provide an AI decision-making process that is understandable through a detailed description of every stage of an application.

5.4.2 Governance and Regulatory Frameworks

Governments are providing rules and regulations for developing an AI application. For example, Europe's GDPR and the European Union's upcoming Artificial Intelligence Act can be thoroughly analysed. GDPR provides detailed guidelines to get consent for data collection and model training, explains how the privacy of users should be protected, and how data minimisation should be achieved. Following Europe and the European Union's rules, other countries are also building up their own rules and regulations for AI firms to follow.

5.4.3 Stakeholder Engagement in AI Development

As we have discussed earlier, there is a huge requirement to involve stakeholders in the life cycle of AI tool and application development. These stakeholders may be industry practitioners, industry veterans, government policymakers, the users of the AI system, researchers or experts in the field. They can create an advisory board where the decisions will be made based on the meetings and discussions. This way, these AI tools can understand the ethical risks based on a particular society and also based on universal standards.

5.4.4 AI Ethics Education and Capacity Building

The mass education on artificial intelligence ethics is helpful in developing a sustainable framework of AI tool development, as it builds the capacity of the masses to understand ethics and its implications while using an AI tool:

Training AI Practitioners and Policymakers: For the AI practitioners, such as engineers, coders, and policymakers, ethics should be part of their technical syllabus and should be taught thoroughly at the college and university level. The future professionals can be enabled with the ethical responsibility in AI development if their engineering syllabus is combined with social sciences, law, and ethical philosophy.

Interdisciplinary Curriculum Development: AI is not limited to any particular domain; rather, every discipline is now being developed using AI. Regarding the requirement of the time, a syllabus which integrates artificial intelligence not only with computer science but also with human rights, ethics, policymaking, and other domains of work is needed. This way, it will enable the professionals to understand the social needs and values while developing the AI applications.

5.5 Public Awareness and Media Literacy

It is required to spread awareness about the AI deployment and its ethical consequences through media and school awareness campaigns. In this way, people can save themselves from the misuse of AI tools and can effectively use AI as a tool for their benefit.

5.6 Case Studies in AI Ethics

The case studies can provide a better understanding of how ethical considerations come into the picture and how careful monitoring and regulation of the ethical considerations can be used to mitigate them and solve the issues [16].

5.6.1 Bias in Criminal Justice Algorithms (COMPAS)

In the US, the COMPAS system is used to understand the possibilities of a criminal repeating the same crime or another crime again. However, due to the bias in the training data, the COMPAS system was prone to flag people from a certain demographic as "high risk". However, their post-parole behaviour actually does not reflect them to be flagged as "high risk". This case study teaches how bias in the underlying dataset can affect the transparency of the system and can have an impact on unfair long-term sentences or denial of payroll.

5.6.2 Privacy Concerns in Facial Recognition Technology

The facial recognition technology is a revolution in artificial intelligence-based tools. From retail markets to law enforcement agencies to traffic rules and regulations departments, they widely use facial recognition systems. Although this technology brings better security in society, it also raises many ethical concerns. First of all, this has created a culture of surveillance. Also, most of the surveillance systems do not have a very high accuracy, which means that they are prone to misidentifying people, especially from communities that are not mainstream. This ethical concern requires a solution in the shape of privacy-protecting mechanisms so that users retain control over their own images and video data, and that should not be sold for any other business benefit or non-consent random surveillance.

5.6.3 Ethical Dilemmas in Autonomous Vehicles

The autonomous vehicles are a benchmark of artificial intelligence with computer vision techniques. Autonomous vehicles make decisions in a fraction of a second and in real time so that they can achieve speed and turns. The ethical complications lie with autonomous vehicles when human safety is involved in the decision-making. The Classic trolley problem explains that autonomous vehicles pose a challenge when it has to choose between two morally perfect options: protect the car and the passenger versus protect the pedestrian on the road crossing. Another key moral concern is regarding the accidents. Who will take the responsibility, the software developer, the manufacturer, or the end user who operates and sits inside the car to get a solution for these ethical dilemmas? We need a closed collaboration among the ethics government policymakers, legal experts, and the manufacturer's side engineers and other experts to get a solution in this regard.

Figure 5.4 illustrates the broader and detailed landscape of AI ethics.

AI Ethics: Challenges and Approaches

Characteristic	Ethical Guidelines & Principles	Governance & Regulatory Frameworks	Stakeholder Engagement	AI Ethics Education & Capacity Building
Focus	Societal welfare, harm avoidance	Ethical compliance, accountability	Diverse participation, societal values	Knowledge dissemination, skill development
Key Elements	Beneficence, non-maleficence, autonomy, justice, explicability	Data protection, transparency, non-discrimination	Interdisciplinary forums, public consultations	Training, curriculum development, public awareness
Examples	Emphasis on societal benefit and fairness	GDPR, EU AI Act, national initiatives	Researchers, practitioners, policymakers, society	Ethics training for developers, public education

Fig. 5.4 Comprehensive framework of AI ethics challenges and approaches

5.7 Future Directions in Ethical AI

There is a need for an ethical framework design and keeping well-documented regulations in line with the pace of AI technology. Current research focuses on evaluating societal impacts, integrating fairness-aware learning, real-time bias mitigation, and participatory design with diverse stakeholders [17]. Developers of AI products should collaborate with engineers, AI veterans, experts, and government policymakers so that social implications and usability of AI are ensured without harming the people. However, the balance in making policy should also be maintained as the progress of AI development can be reduced due to strict regulations [18]. Also, focusing on the safety and accuracy of AI models is critically required during AI model development. The safety of AI models is to protect them from malicious attacks by unethical hackers [19]. The high accuracy of an AI model provides reliability and enhances trust for individuals. It is also important to maintain the transparency of an AI model using explainable AI-related features. Expendable AI will clearly report how each component of an AI framework is creating a certain output. This will enhance the trustworthiness of AI use in critical domains like healthcare and legal systems.

5.8 Conclusion

The domain of artificial intelligence and the philosophical concept of ethics bring challenges and opportunities for learners of computer science. AI is penetrating every domain of work, and this is bringing the requirement of making it always free from unethical practices, a highly required expectation. This chapter is designed to first introduce its readers to the technical foundation of artificial intelligence concepts, then provide an introduction to the ethical concerns associated with artificial intelligence tools that often occur regularly in society, and then discuss the framework to guide ethical artificial intelligence tool development.

The main theme of this chapter lies in the fact that although the benefits of AI are surprising in every field of work, at the same time, the ethical risks are also penetrating everywhere. Ethical AI-enhancing practices, such as fairness-aware algorithms and explainable AI (XAI), should work collaboratively. They should also strictly follow the government framework of regulations and actively engage engineers, specialists, ethicists, and end users.

Ethical AI isn't just a technical issue—it's a responsibility we all share as a society. To truly build trust, protect human rights, and ensure fairness in a world shaped by intelligent machines, we need more than just powerful technology—we need values, accountability, and care guiding every decision. This journey calls on developers, policymakers, and communities to work together to create systems that reflect our highest ethical standards.

Moving forward, real progress will depend on open dialogue, ongoing research, and flexible regulations that keep pace with AI's rapid growth. The challenges aren't limited to engineering; they touch on social and ethical questions too. That's why we must focus on fairness, transparency, and strong governance that holds AI systems—and their creators—accountable. By confronting biases, making AI more understandable, and ensuring diverse voices are included in the conversation, we can shape a future where technology doesn't just advance—but advances for everyone, equitably and responsibly.

References

1. Hassan MAM, Sardar TH, Fahim MFH, Mohamed MS, Suleyman RM, Usman MM. Artificial intelligence and deep learning applications on big data computing frameworks. In: Big data computing. CRC Press;2024. pp. 196–211.
2. Sardar TH, Das S, Pandey BK. Explainable AI (XAI): Concepts and theory. In: Medical data analysis and processing using explainable artificial intelligence. CRC Press;2023. pp. 1–18.
3. Mian S. Foundations of artificial intelligence and applications. J Artif Intell Technol. 2022;2(1):1–2.
4. Bhattacharjee R, Sasi SD, Thakran S. Artificial intelligence, its components, and crucial technologies for its implementation. In: Radiomics and radiogenomics in neuro-oncology. Academic Press;2024. pp. 23–36.
5. Swasthika Jain TJ, Sardar TH, Sammeda Jain TJ, Guru Prasad MS, Naresh P. Facial expression analysis for efficient disease classification in sheep using a 3NM-CTA and LIFA-based framework. IETE J Res. 2025; 1–15. https://doi.org/10.1080/03772063.2025.2498610
6. Sardar TH, Ansari ZA, Theerthagiri P, Karthikeyan P, Ayyasamy V, Saini DKJB. MapReduce-enhanced fuzzy K-least medians for qualitative clustering of document big data. Concurr Comput: Pract Exp. 2025;37(4–5): e70035.
7. Sardar TH, Ansari Z. Distributed big data clustering using MapReduce-based fuzzy C-medoids. J Inst Eng (India): Ser B. 2022;103(1):73–82.

8. Rajan K, Saffiotti A. Towards a science of integrated AI and Robotics. Artif Intell. 2017;1(247):1–9.
9. Wertheimer M. A Gestalt perspective on computer simulations of cognitive processes. Comput Hum Behav. 1985;1(1):19–33.
10. De Silva D, Alahakoon D. An artificial intelligence life cycle: from conception to production. Patterns. 2022; 10:3(6).
11. Baeza-Yates R. Ethical challenges in AI. In: Proceedings of the fifteenth ACM international conference on web search and data mining;2022. pp. 1–2.
12. Toy T. Transparency in AI. AI Soc. 2024;39(6):2841–51.
13. Zhang B. Accountability and responsibility for AI-enabled conduct. In: Research handbook on warfare and artificial intelligence. Edward Elgar Publishing;2024. pp. 216–33.
14. Mazurek G, Małagocka K. Perception of privacy and data protection in the context of the development of artificial intelligence. J Manag Anal. 2019;6(4):344–64.
15. Ghallab M. Responsible AI: requirements and challenges. AI Perspect. 2019;1(1):1–7.
16. Vakkuri V, Kemell KK, Abrahamsson P. Implementing ethics in AI: initial results of an industrial multiple case study. In: International conference on product-focused software process improvement. Cham: Springer International Publishing;2019. pp. 331–8.
17. Firmansyah G, Bansal S, Walawalkar AM, Kumar S, Chattopadhyay S. The future of ethical AI. In: Challenges in large language model development and AI ethics. IGI Global;2024. pp. 145–77.
18. Mager A, Eitenberger M, Winter J, Prainsack B, Wendehorst C, Arora P. Situated ethics: ethical accountability of local perspectives in global AI ethics. Media Cult Soc. 2025;47(5):1028–41.
19. Wu T, Dong Y, Dong Z, Singa A, Chen X, Zhang Y. Testing artificial intelligence system towards safety and robustness: state of the art. IAENG Int J Comput Sci. 2020;47(3):449–62.

Chapter 6
Big Data and Data Science

6.1 Introduction

The digital revolution has created an unprecedented era of data proliferation, fundamentally altering how organisations operate and create value. The convergence of ubiquitous computing, Internet connectivity, and sensor networks has resulted in exponential increases in data generation rates.

6.1.1 *Definition, Emergence, and Importance of Big Data*

Big data refers to datasets whose size, complexity, and generation velocity exceed traditional database management capabilities. Its emergence stems from widespread digital technology adoption, IoT device proliferation, social media growth, and business process digitisation [1].

Big data's importance extends beyond technological considerations. Data-driven organisations demonstrate 23 times higher customer acquisition rates, 6 times better retention, and 19 times greater profitability according to McKinsey research. The transformative impact enables personalised customer experiences, predictive maintenance, real-time optimisation, scientific breakthroughs, and evidence-based policymaking [2].

6.1.2 *The "V's" of Big Data*

The characteristics of big data, which separate it from traditional data, are presented by the "v's" of big data [3–7].

T. H. Sardar, *The Spectrum of Computer Science: Emerging Technologies and Trends*, Studies in Smart Technologies, https://doi.org/10.1007/978-3-032-19632-3_6

Volume represents data scale, measured in petabytes or exabytes. IDC predicts global datasphere growth from 33 zettabytes in 2018 to 175 zettabytes by 2025.

Variety encompasses heterogeneous formats including structured, semi-structured (XML, JSON), unstructured (text, images, videos), and streaming data, necessitating flexible storage architectures and adaptable processing frameworks.

Velocity refers to generation, transmission, and processing speed requirements. Real-time applications like high-frequency trading, fraud detection, and recommendation engines require millisecond response times.

Veracity addresses data quality and trustworthiness challenges, including inconsistencies, errors, missing values, and biased sampling requiring robust governance frameworks and validation procedures.

Value represents the ultimate objective: extracting actionable insights and measurable benefits. Value creation requires technical capabilities, domain expertise, and organisational readiness to act upon insights.

The big data's 5 Vs pyramid, as illustrated in Fig. 6.1, provides the foundational layers of Volume, Velocity, Variety, Veracity, and Value, emphasising the sheer quantity of data, processing speed, diverse formats, accuracy, and potential insights. This hierarchical model highlights how each layer builds upon the previous to maximise data utility in various industries.

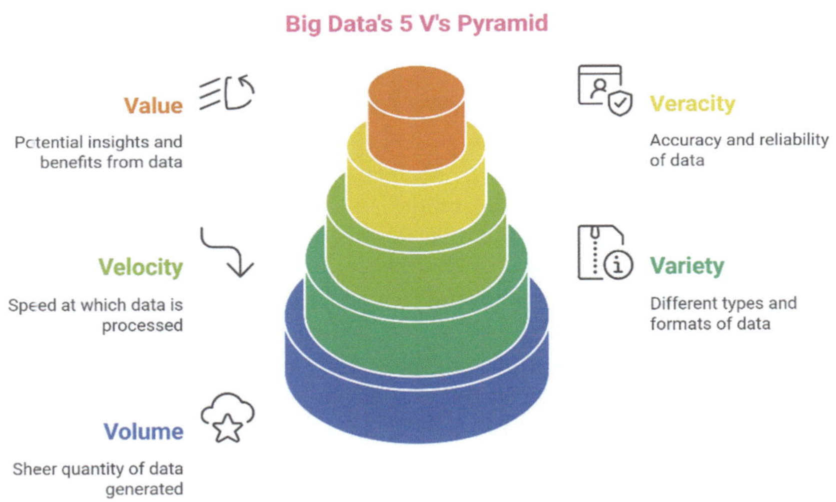

Fig. 6.1 Hierarchical representation of big data's 5 Vs

6.2 Big Data Ecosystem and Applications

6.2.1 Sector-Wise Applications

Big data has penetrated its impact in almost all key sectors [8].

Healthcare applications include predictive care models using machine learning to identify at-risk patients, personalised medicine leveraging genomic data and treatment histories, and medical imaging analysis. IBM Watson for Oncology analyses medical literature and patient data for evidence-based treatment recommendations.

Finance focuses on risk management, fraud detection, and algorithmic trading. Machine learning algorithms analyse transaction patterns for real-time fraud detection, while high-frequency trading systems execute trades within microseconds using market data analysis.

Retail and E-commerce leverage personalised recommendations, dynamic pricing, and inventory optimisation. Amazon's recommendation system generates 35% of revenue through personalised suggestions based on customer behaviour analysis.

Science and Research spans astronomy, climate science, and social research. The Large Hadron Collider generates 50 petabytes annually, while climate researchers utilise satellite imagery and sensor data for environmental modelling.

Government and Public Services implement smart city initiatives, predictive policing, and evidence-based policymaking through integrated sensor data, demographic analysis, and resource optimisation.

Transforming industries with big data showcases, as in Fig. 6.2, its impact across government/public services, healthcare, science and research, finance, and retail, with a central data cloud driving innovation. Each sector leverages big data to enhance services, research, and operational efficiency.

6.2.2 Case Study: Netflix's Data-Driven Content Strategy

Netflix collects over 200 billion daily events from 200+ million subscribers, including viewing patterns, pause points, completion rates, and search queries. The recommendation algorithm uses collaborative filtering, content-based filtering, and deep learning for personalisation, generating 80% of watched content through algorithmic suggestions.

Beyond recommendations, Netflix analyses viewing patterns and demographic data for original content development. The "House of Cards" decision was partly based on data indicating demand for political dramas. This data-driven approach enables successful content creation while minimising financial risks.

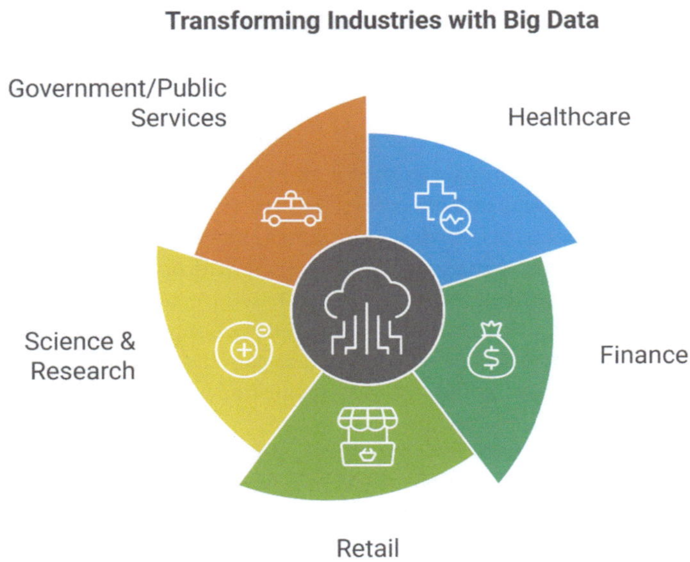

Fig. 6.2 Big data's transformative role across industries

6.2.3 Case Study: Precision Agriculture and IoT Data Analytics

Precision agriculture integrates IoT sensors monitoring soil moisture, nutrients, temperature, and crop growth with satellite imagery and weather data. John Deere's platform combines GPS-equipped tractors, soil sensors, and satellite imagery for optimised planting, fertilisation, and irrigation decisions.

Machine learning algorithms predict optimal planting dates and crop varieties, while real-time sensor data adjusts irrigation based on soil moisture and weather forecasts. Variable rate technology enables precise fertiliser application based on localised conditions. Farmers report 10–20% yield improvements and 15–25% input cost reductions.

6.2.4 Competitive Advantage Through Big Data

Data-driven decision-making enables accurate market predictions, customer insights, and operational optimisations. Predictive analytics support proactive strategies, while personalisation improves customer experiences. However, competitive advantage requires data literacy, robust governance, and an experimentation culture beyond technological implementation.

6.3 Data Processing Frameworks and Techniques

6.3.1 Data Collection and Storage

Hadoop Distributed File System (HDFS) provides fault-tolerant storage across commodity hardware, dividing files into blocks (128 MB–256 MB) replicated across multiple nodes. Recent enhancements include HDFS Federation and High Availability configurations [9–11].

NoSQL Databases address relational database limitations through flexible schemas and massive scalability. Document databases (MongoDB) store JSON-like documents, key-value stores (DynamoDB) provide high-performance access, column-family databases (HBase) handle sparse data, and graph databases (Neo4j) manage relationships [12].

Polyglot Persistence combines multiple database technologies optimising for specific use cases, such as relational databases for transactions, document databases for catalogues, and graph databases for recommendations [13].

Data Lakes provide centralised repositories storing structured and unstructured data without predefined schemas, supporting schema-on-read approaches and integration with various analytics tools [14].

As Fig. 6.3 shows, the Layered Data Architecture outlines a structured approach with Data Sources at the base, followed by Storage, Processing, and Access layers, optimising data management. This model ensures efficient handling and utilisation of diverse data formats for actionable insights.

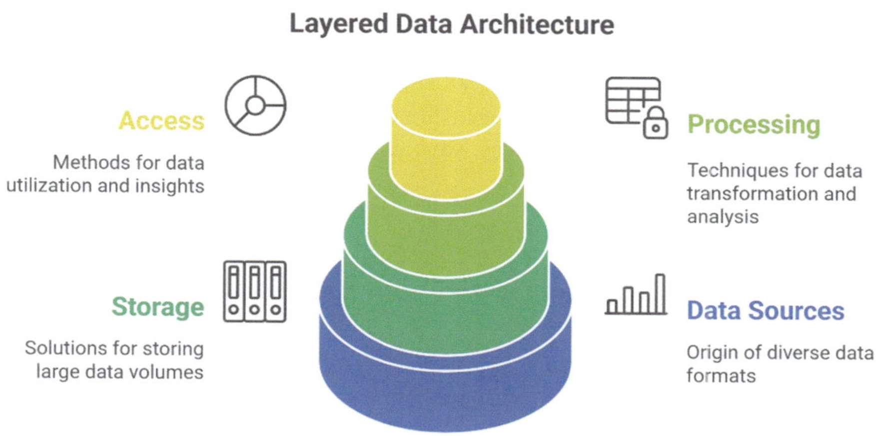

Fig. 6.3 Structured layers of data architecture

6.3.2 Data Cleaning and Preprocessing

Data Profiling systematically examines data structure, content, and quality, identifying data types, distributions, null rates, and relationships to design appropriate cleaning strategies.

Missing Data Handling includes deletion methods and imputation techniques like mean/median replacement, regression imputation, and multiple imputation accounting for uncertainty.

Outlier Detection and Treatment uses statistical methods (Z-score, IQR) and machine learning approaches (isolation forests) to identify unusual observations, with treatment strategies including correction, removal, or transformation.

Standardisation and Normalisation ensure variables with different scales can be meaningfully compared through Z-score normalisation or min–max scaling, plus categorical encoding and feature engineering [15].

6.3.3 Data Analytics

Descriptive Analytics summarises historical data through statistical measures and visualisations, providing insights into past performance and trends via business intelligence dashboards and reporting.

Diagnostic Analytics investigates why events occurred through correlation analysis, hypothesis testing, and regression analysis to provide explanatory understanding.

Predictive Analytics forecasts future events using statistical and machine learning models, including time series analysis (ARIMA), classification algorithms, and ensemble methods for proactive decision-making.

Prescriptive Analytics recommends optimal actions combining predictive insights with optimisation techniques, simulation, and reinforcement learning for decision support [16].

6.3.4 Data Visualisation and Decision Support

Interactive Dashboards present key metrics through visually intuitive formats enabling rapid comprehension. Modern platforms like Tableau and Power BI provide drag-and-drop interfaces with real-time data connections.

Exploratory Visualisation supports analytical processes through scatter plots, histograms, and heat maps, while advanced techniques like parallel coordinates reveal high-dimensional patterns.

Narrative Visualisation combines insights with storytelling techniques, sequencing visualisations to build compelling arguments for diverse audiences.

Automated Insight Generation leverages machine learning to identify patterns and generate natural language descriptions, democratising data analysis capabilities [17].

6.4 Tools and Platforms for Big Data Science

6.4.1 Hadoop Ecosystem

HDFS provides distributed storage foundation with fault tolerance across commodity hardware clusters through block replication and automatic failure handling.

MapReduce enables distributed computation through map and reduce functions, though its batch-oriented approach and disc-based storage limit real-time applications.

Apache Hive provides SQL-like querying capabilities, translating HiveQL to MapReduce, Tez, or Spark jobs for business analyst accessibility.

Apache Pig offers high-level scripting through Pig Latin for data transformation workflows, particularly excelling at ETL operations.

6.4.2 Apache Spark

Spark Core provides distributed computing through Resilient Distributed Datasets (RDDs) with in-memory processing and lazy evaluation for performance optimisation.

Spark SQL enables structured data processing with familiar SQL syntax, featuring the Catalyst optimiser and Tungsten execution engine for performance improvements.

MLlib delivers scalable machine learning algorithms for classification, regression, clustering, and collaborative filtering with pipeline APIs for complex workflows.

GraphX provides graph processing capabilities for network analysis tasks like PageRank and connected components.

Spark Streaming/Structured Streaming enables real-time processing with micro-batch and continuous models, treating streaming data as continuously updating tables.

6.4.3 *Jupyter and Interactive Environments*

Jupyter Notebooks provide web-based interactive computing combining code execution, visualisation, and documentation with multi-language support through kernels.

JupyterHub enables multi-user environments with authentication, resource management, and scalability through Kubernetes integration.

Alternative platforms include Apache Zeppelin with enhanced collaboration and Google Colab with free GPU access for machine learning experiments.

6.4.4 *Cloud-Based Solutions*

Amazon Web Services offers EMR for managed clusters, Redshift for warehousing, Kinesis for streaming, and SageMaker for machine learning, with S3 providing central data repository and extensive service integration.

Microsoft Azure provides HDInsight, Synapse Analytics, Stream Analytics, and Machine Learning with tight Office 365 integration and Power BI reporting.

Google Cloud Platform provides a set of tools for different services. Instead of owning the hardware and management staff for data warehousing, Google Cloud provides it through Dataproc and BigQuery tools. It also provides the Dataflow tool for artificial intelligence-based developments and for the best processing of big data in a distributed cloud network.

6.5 Challenges in Big Data and Data Science

6.5.1 *Data Privacy and Security*

Big data has a unique feature called "Value" as it is useful for monetary and governance benefits. This "value" of big data makes it prone to theft, misuse, and at risk from malicious web robot access. It is required to ensure that the big data is safe from misuse and should be used in accordance with ethical considerations only.

6.5.1.1 Data Anonymisation

Data anonymisation is a technique where personal details are removed from the content to protect the privacy of the author. However, this job is difficult to achieve because advanced searches can provide other information regarding the available similar content of the author, which later helps in retrieving the author's information.

It has been a big issue to keep the balance between getting the content into its place while hiding the identity of the authorship.

6.5.1.2 Security Architecture for Big Data

Big data systems are complex and distributed across many machines or cloud services. Designing a strong security architecture that covers all entry points, prevents cyberattacks, and ensures safe data storage and processing is a tough task. Any weak spot in the system can become a major risk.

6.5.1.3 Encryption Strategies

For data protection, data encryption algorithms are a widely used technique. However, this data encryption technique carries with it extra processing time. Big data is already voluminous, and adding extra processing overweight is also always targeted to be reduced as much as possible. The engineers have to choose a suitable data encryption technique while considering the low processing overhead. Also, safety in managing distributed keys in a large big data distributed environment is also a concern.

6.5.1.4 Access Control and Audit

In a large enterprise, different stakeholders require access to data for a certain amount of time. It is challenging to maintain an audit log which ensures that a particular authorised stakeholder is accessing the data for a specific time.

Figure 6.4 provides a visual representation of the comprehensive data privacy and security framework.

6.5.2 Ethical and Regulatory Concerns

Data science has emerged as one of the major computer science subdomains of the twenty-first century. Data science technique deployment with big data raises concerns regarding ethical complications, such as privacy-preserving data collection, user rights maintenance, data protection, algorithmic accountability, and fairness of its results.

6.5.2.1 General Data Protection Regulation (GDPR)

In 2018, the European Union 2018 created the General Data Protection Regulations. It has been created to maintain personal privacy and the protection of data through

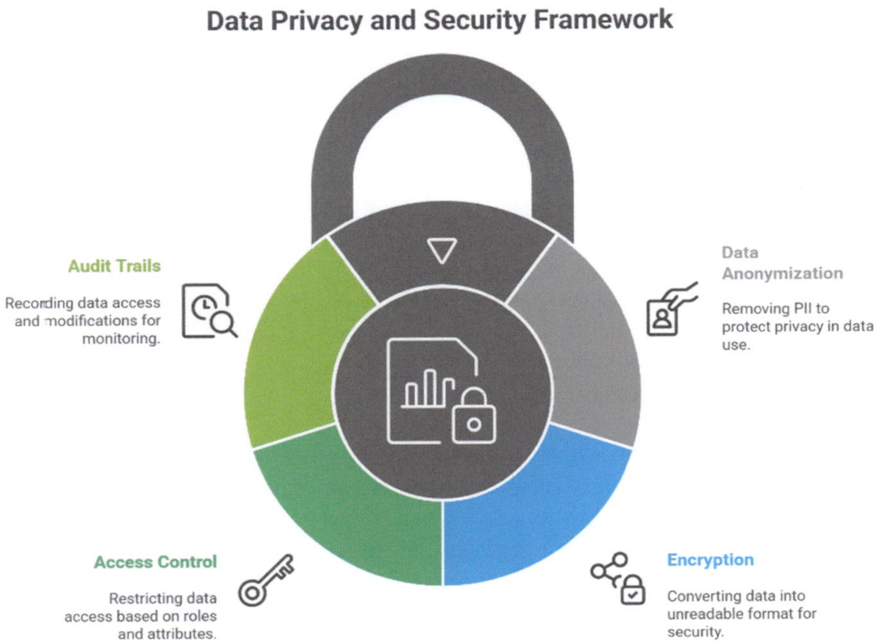

Fig. 6.4 Comprehensive data privacy and security framework

rules and regulations. The invitation suggests processing data sources only with required consent, maintaining data minimisation and auditing to keep accountability of the data. The data science practices which follow GDP and regulations will reduce legal risks, individual autonomy, and the integrity of data.

6.5.2.2 Individual Rights and Data Management

Providing individual rights is the core ethical part of data management, where we specify user rights in accessing, deleting, finding, and restricting the processing of a user's personal data. Data science applications, while collecting the data and processing, should comply with this ethical standard. The big data systems must ensure the development of methods for consent management, and secure the storage from data theft and misuse.

6.5.2.3 Health Insurance Portability and Accountability Act (HIPAA)

The health-related data is a sensitive domain of privacy and security management. The United States has a governing body called Health Insurance Portability and Accountability (HIPAA). HIPAA consists of directions and regulations for preventing the misuse of healthcare data. Data science and big data-based healthcare tools should follow the compliance provided by HIPAA to ensure data protection and trust building.

6.5.2.4 Algorithmic Bias and Fairness

The algorithmic bias is a result of training data science techniques and algorithms based on a skewed dataset with biases and assumptions. The demographic, religion-based, region-based, and ethnicity-based biases in the training dataset affect decision-making and inadvertently affect sensitive domains like credit scoring, justice, and recruitment.

6.5.3 Scalability and Cost Challenges

6.5.3.1 Infrastructure Scaling

The vertical scaling, or scaling of processing capacity in terms of hardware specification enhancement in a computing system, is obsolete. Horizontal scaling, while adding more available hardware together, is an easy way of attaining higher capacity of storage and processing power. But this introduces the issue of coordination and consistency, solving the issue of scalability at a low expense. As a solution, a cloud-based infrastructure scaling technique can provide temporary solutions but create integration complexity.

6.5.3.2 Data Storage Costs

Big data storage, as it needs exponential data growth over time, requires a huge space. However, as a solution, data lifecycle management techniques can be used. They reduce the expenses using different storage classes, compression of data, and intelligent tiering.

6.5.3.3 Computational Cost Optimisation

Computational cost can be optimised if we use specific tools for suitable applications. Similarly, workload scheduling techniques choose resources based on their availability, prices, and performance for the total cost of the project.

6.5.3.4 Performance Optimisation

Optimised performance requires proper coordination of computational systems, storage systems, and the networks. Recovery optimisation techniques, such as predicate pushdown techniques, usually increase analytical query performance, and a cache-based storage system enhances the performance of memory access.

6.5.4 Skills and Workforce Gaps

6.5.4.1 Data Science Skills Shortage

As the data science field is developing very fast in every domain, we need experts specialising in the domain. For example, in statistical analysis, machine learning, and programming, working domain-specific expertise is shorter than the required. The traditional education does not sufficiently provide this expertise to the students, while online platforms are actually providing modern tools which has to be used by experts only.

6.5.4.2 Organisational Change Management

The management of the traditional organisations is not aware or capable of changing based on the need for data science for big data. It is needed to ensure the cultural resistance to data-driven approaches using communication, programs, and training for the existing staff.

6.5.4.3 Continuous Learning Requirements

The existing kill force workers should be equipped with data science-related knowledge and skills. Continuous learning skills can be employed by the organisations so that the existing workers can enhance their skills with new technology. This can be promoted via technical clubs and different competitive price-based events.

6.5.4.4 Cross-Functional Collaboration

Collaboration has become easier in today's connected world. Cross-functional collaborative teams should be managed with clear rules and responsibilities, and should be connected using regular meetings and encouraged by performance-based incentives.

6.6 Emerging Trends and Research Directions

6.6.1 AI and Big Data Synergy

6.6.1.1 Deep Learning at Scale

Deep learning has become a key technique of machine learning. It can process big data through huge computational power. Deep learning distributed techniques and tools, such as Horovod, can provide learning through billions of parameters while achieving optimised memory-based operations.

6.6.1.2 Transformer Architectures

Natural language processing is a technique that automatically reads human language text, processes it for understanding, replying, or generating natural language. Transformer architectures have revolutionised natural language processing. Models, such as GPTs, have path-breaking achievements in transformer-based natural language processing.

6.6.1.3 Computer Vision Applications

Computer vision applications require training on image data from CCTV cameras, high-resolution videos from autonomous vehicles, and medical image data. This very large and rapidly generating data has to be accessed in real time in lesser time so that autonomous vehicles can react in a fraction of a second and medical imaging can provide better diagnosis earlier.

6.6.1.4 Reinforcement Learning at Scale

To make reinforcement learning at scale, we must generate a huge amount of experience data and train on the data. A parallel execution, with optimisation and multi-agent learning, can be provided using distributed frameworks.

6.6.2 *Edge and Fog Computing*

6.6.2.1 Edge Computing

Edge computing brings capabilities closer to data sources, reducing latency and bandwidth while improving privacy. IoT devices generate data overwhelming network infrastructure if centrally transmitted, enabling local processing and filtering.

6.6.2.2 Edge AI and Federated Learning

Federated learning enables machine learning across distributed datasets without centralisation, addressing privacy concerns. Edge AI deploys models directly on devices enabling real-time inference without connectivity through model optimisation techniques.

6.6.2.3 Fog Computing

Fog computing extends cloud capabilities through intermediate processing nodes providing more resources than edge devices while remaining distributed. This hierarchical approach enables flexible resource allocation and improved performance.

6.6.2.4 Industrial IoT Applications

Industrial IoT combines massive sensor deployments with real-time processing requirements. Predictive maintenance analyses sensor data predicting failures, while smart grid applications require real-time electrical consumption analysis for optimisation and stability.

6.6.3 *Quantum Computing and Big Data*

6.6.3.1 Quantum Algorithms for Data Analysis

Quantum computing provides a new Paradise of quantum bits for computing, which has proven to provide exceptional time performance for linear algebra, search problems and optimisation techniques. Although the hardware restrictions and limitations are a major issue, machine learning algorithms are not efficient for pattern recognition techniques.

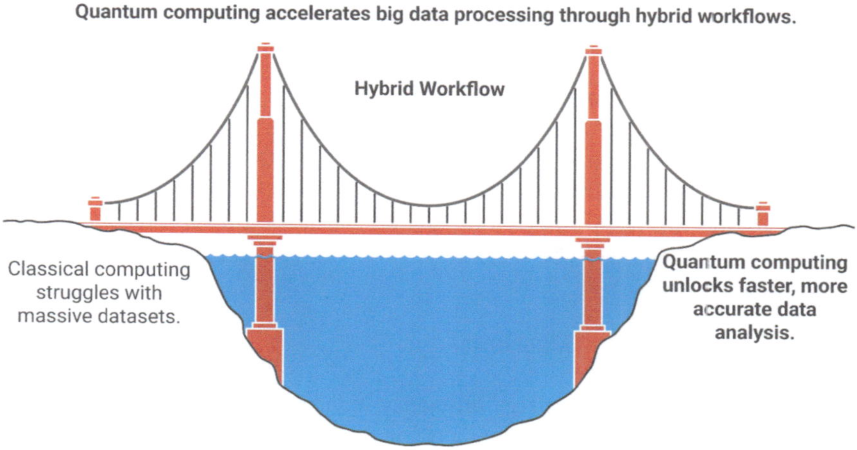

Fig. 6.5 Hybrid workflow of quantum and classical computing

6.6.3.2 Quantum Database Search

Quantum algorithms such as Grover's algorithm have the potential for big data query optimisation, for pattern matching and unstructured database search. The quantum optimised traditional algorithms can even perform better using a classical hardware system for clustering and recommendation jobs.

6.6.3.3 Hybrid Quantum–Classical Systems

Many quantum-based applications are developed in a traditional way due to practical constraints, such as the high price of quantum computing systems, which are now combining quantum-based algorithms with classical computing systems. Proven effective for addressing the quantum error connection challenge, while providing limited coherence.

Figure 6.5 shows how quantum computing accelerates big data processing through a hybrid workflow.

6.6.3.4 Current Hardware Limitations

Current quantum computers suffer high error rates, limited coherence times, and restricted qubits. Quantum error correction requires hundreds of physical qubits per logical qubit, severely limiting effective computational capacity.

6.6.4 *Sustainable Data Science*

6.6.4.1 Green AI Initiatives

Environmental impact receives increasing attention as computational demands grow exponentially. Training GPT-3 generated carbon emissions equivalent to 700,000 miles of driving. Green AI focuses on efficient algorithms, renewable energy, and optimised resource utilisation.

6.6.4.2 Energy-Efficient Algorithms

Algorithm design considers energy efficiency alongside performance. Techniques such as efficient architectures, early stopping, and adaptive resource management are used in this context.

6.6.4.3 Carbon-Aware Computing

The carbon-aware computing uses scheduling-based load specification, temporal load shifting, and geographic shifting to consider the electricity source carbon intensity.

6.6.4.4 Efficient Data Centre Design

There are many techniques which are implemented for the efficient design of a data centre. Examples are advanced cooling mechanisms, optimisation of servers, and renewable energy source integrations which are key efforts in this domain.

6.6.4.5 Sustainable Data Practices

Sustainable data practices refer to considering lifecycle management practices to reduce environmental impact while enhancing storage management, such that minimal data required for data analytics should be stored only.

6.7 Conclusion

6.7.1 Summary of Contributions

This chapter provided a comprehensive examination of big data and data science, covering fundamental concepts, technological frameworks, practical applications, and emerging challenges. The features of big data are provided in this chapter in terms of the "V's" framework, which goes beyond Volume, Variety, Velocity, Veracity, and Value. Sector-specific applications show how big data impacts the key sectors such as healthcare, finance, retail, science, and government.

This chapter shows how traditional databases are now shifting towards sophisticated distributed architectures for efficient data processing. It shows how cloud computing architectures and frameworks have made big data-based data science application development less expensive, available, and easier to collaborate on.

6.7.2 Open Research Problems

One of the key research problems in data science and the big data domain lies in preserving the privacy of the data users while generating the best data science solution. Current approaches impose performance penalties, limiting applications. Automated insight generation must bridge statistical pattern detection and meaningful business insight gaps.

Real-time streaming analytics continues to challenge frameworks for complex operations requiring historical context. Ethical AI and algorithmic fairness require ongoing research, developing metrics, methodologies, and governance frameworks, ensuring responsible technology use.

6.7.3 Future Outlook

Future big data and data science will be shaped by converging trends. AI integration will enable sophisticated analytical approaches while creating explainability challenges. Edge computing will bring capabilities closer to sources, enabling real-time applications while addressing privacy concerns.

Quantum computing may provide exponential speedups for certain problems, though practical applications remain years away. Sustainability considerations will increasingly influence architecture decisions as organisations balance computational capabilities with environmental responsibility.

Democratisation through automated machine learning will expand analytical capabilities beyond traditional roles, requiring new education, governance, and quality assurance approaches. Regulatory developments will continue shaping practices as governments develop data protection and algorithmic accountability frameworks.

Success requires maintaining technological agility while developing foundational capabilities in data governance, analytical thinking, and ethical decision-making. The future belongs to those harnessing big data power while navigating challenges responsibly and effectively.

References

1. Sardar TH, Ansari Z. Distributed big data clustering using MapReduce-based fuzzy C-medoids. J Inst Eng (India): Ser B. 2022; 103(1):73–82.
2. Sardar TH, Ansari Z. MapReduce-based fuzzy C-means algorithm for distributed document clustering. J Inst Eng (India): Ser B. 2022; 103(1):131–42.
3. Sardar TH, Ansari Z. Partition based clustering of large datasets using MapReduce framework: an analysis of recent themes and directions. Futur Comput Inform J. 2018;3(2):247–61.
4. Sardar TH, Muttineni A, Ranjan R. (2024). The future of big data in customer experience and inventory management. In: Big Data Computing. CRC Press; pp. 233–48.
5. Hassan MAM, Sardar TH, Fahim MFH, Mohamed MS, Suleyman RM, Usman MM. Artificial intelligence and deep learning applications on big data computing frameworks. In: Big data computing. CRC Press;2024. pp. 196–211.
6. Sardar TH, Pandey BK (Eds). Big data computing: advances in technologies, methodologies, and applications. CRC Press;2024.
7. Sardar TH, Ansari ZA, Theerthagiri P, Karthikeyan P, Ayyasamy V, Saini DKJB. MapReduce-enhanced fuzzy K-least medians for qualitative clustering of document big data. Concurr Comput: Pract Exp. 2025;37(4–5): e70035.
8. Anwar MJ, Gill AQ, Hussain FK, Imran M. Secure big data ecosystem architecture: challenges and solutions. EURASIP J Wirel Commun Netw. 2021;2021(1):130.
9. Ansari Z, Afzal A, Sardar TH. Data categorization using Hadoop MapReduce-based parallel K-means clustering. J Inst Eng (India): Ser B. 2019;100(2):95–103.
10. Sardar TH, Ansari Z, Khatun A. An evaluation of Hadoop cluster efficiency in document clustering using parallel K-means. In: 2017 IEEE international conference on circuits and systems (ICCS). IEEE;2017. pp. 17–20.
11. Sardar TH, Faizabadi AR, Ansari Z. An analysis of data processing using Mapreduce paradigm on the Hadoop framework. Int J Emerg Res Manag Technol. 2017;6(5).
12. Tripathi N. NoSQL database education: a review of models, tools and teaching methods. J Syst Softw. 2025;25: 112391.
13. Sharma S, Dubey RK, Khatri P. Review of the literature on big data and performance analysis of multimodel and polyglot persistence. In: 2024 international conference on advances in computing research on science engineering and technology (ACROSET). IEEE;2024. pp. 1–6.
14. Zhao X, Zhang C, Guan S. A data lake-based security transmission and storage scheme for streaming big data. Clust Comput. 2024;27(4):4741–55.
15. Sardar TH, Sarkar R, Ahmed SJ, Bandyopadhyay A. A novel ensemble methodology to validate fuzzy clusters of big data. In: Proceedings of the fourth international conference on

trends in computational and cognitive engineering: TCCE 2022. Singapore: Springer Nature Singapore;2023. pp. 267–78.
16. Sardar TH. Reflecting on a decade of evolution: MapReduce-based advances in partitioning-based, hierarchical-based, and density-based clustering (2013–2023). Wiley interdisciplinary reviews: data mining and knowledge discovery;2024. e1566.
17. Alves CMO, Cota MP. Visualization on decision support systems models: literature overview. In: Rocha Á, Adeli H, Reis LP, Costanzo S. editors. Trends and advances in information systems and technologies. WorldCIST'18 2018. Advances in intelligent systems and computing, vol. 745. Springer, Cham;2018. https://doi.org/10.1007/978-3-319-77703-0_73

Chapter 7
Blockchain and Cryptocurrencies

7.1 Basics of Blockchain Technology

7.1.1 Definition and Overview

Blockchain technology utilises cryptography, verification, and a network-wide distributed consensus to introduce a new approach to information storage and transaction management [1]. Based on the principles of cryptography, a blockchain-based distributed ledger system can effectively maintain massive amounts of data over time, incrementally adding blocks of records while maintaining required security [2]. Each block of records is connected using a fixed chain of cryptographic hashing-based technique, which is the most secure technique. This can be achieved while providing security, transparency, and decentralisation.

One of the key contributions of blockchain lies in providing security and trust in a less secure system. Traditional digital transactions often require the involvement of trusted third parties, such as banks or certification authorities. Blockchain technology can replace these trusted third parties and their centralised systems using a distributed consensus mechanism based on algorithms. This way, blockchain enables peer-to-peer transactions.

Figure 7.1 illustrates the major application areas of blockchain, such as across digital identity, supply chain management, voting systems, and healthcare.

7.1.2 How Blockchain Works

A successful blockchain transaction requires interconnected processes to work in coordination [3]. The specifications below provide an overview of this process.

T. H. Sardar, *The Spectrum of Computer Science: Emerging Technologies and Trends*,
Studies in Smart Technologies, https://doi.org/10.1007/978-3-032-19632-3_7

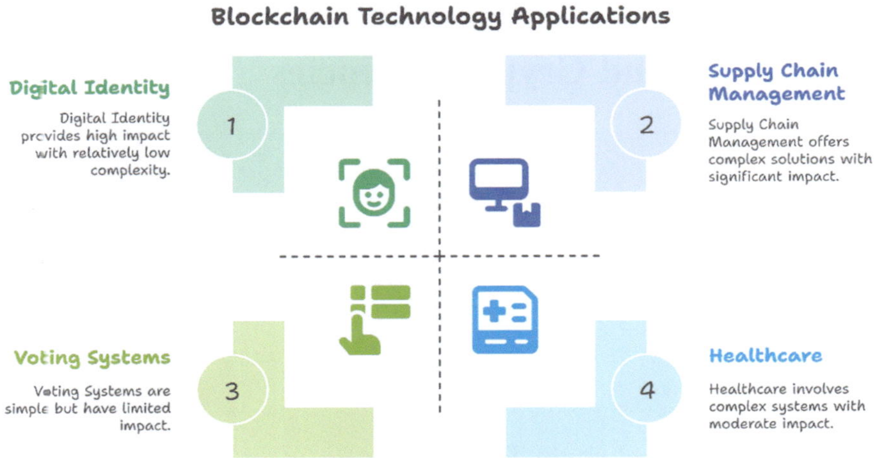

Fig. 7.1 Blockchain's key technology applications

Transaction Initiation: Transactions are initially initiated by users. For example, users can initiate cryptocurrency transaction transfers, which are authenticated by digital signatures and cryptography techniques.

Transaction Broadcasting: Next, the transaction will be broadcast by distributing across the peer-to-peer networks. Network notes will validate the signed transactions in accordance with the rules and regulations. Invalid transactions will be rejected at this phase.

Block Formation: Block formation is performed by specialised nodes called minors or validators. They are generated by collecting valid transactions, adding headers with timestamps, and performing Merkle tree summarisation.

Consensus Achievement: Network participants agree on blockchain's current state through various consensus mechanisms, ensuring distributed agreement despite potentially anonymous and adversarial participants.

Block Addition and Distribution: Newly accepted blocks propagate throughout the network, enabling all participants to update their local blockchain copies.

Figure 7.2 shows the step-by-step process of blockchain operation, beginning with transaction initiation, followed by broadcasting, block formation, consensus, and block addition to the ledger.

7.1.3 Types of Blockchains

The blockchains can be broadly categorised in the below segments [4]:

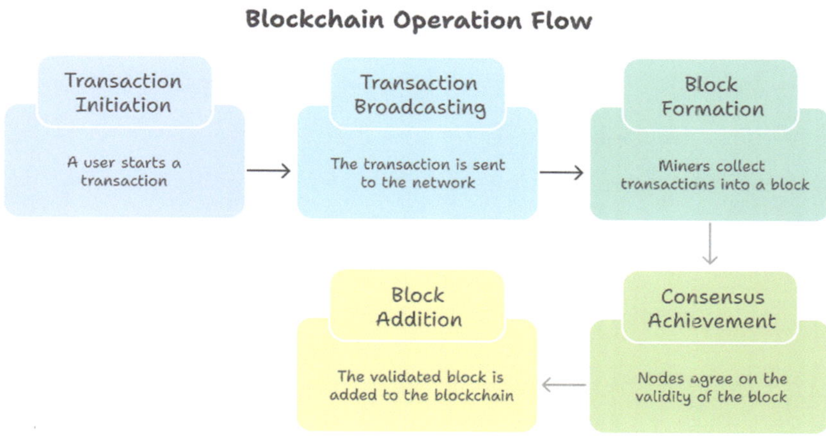

Fig. 7.2 Blockchain operation flow from transaction initiation to consensus achievement

Public blockchains are fully decentralised and permissionless, allowing anyone to join, validate transactions, and participate in consensus. They prioritise transparency, immutability, and censorship resistance. Notable examples include Bitcoin, which emphasises security through proof of work, and Ethereum, which extends blockchain functionality with smart contracts. However, these networks often face challenges of scalability, high energy consumption, and slower transaction speeds, creating a trade-off between decentralisation and efficiency.

Private blockchains operate under restricted access, typically controlled by a single organisation or authority. **Private blockchains** are suitable for large companies in various applications, such as supply chain management, financial statement auditing, and healthcare data records. Private blockchains provide high support and reduce latency with customisable features.

Figure 7.3 compares different blockchain types based on different characteristics and use cases.

Consortium blockchains are semi-decentralised blockchain systems that work with shared governance and reduce centralisation of power in a single data node. They are widely used in banking consortia (e.g. R3 Corda), trade finance, and cross-institutional collaborations, offering both trust and performance in controlled ecosystems.

Hybrid blockchains integrate features of both public and private systems. They enable organisations to maintain privacy for sensitive internal data while selectively sharing specific information with a public blockchain for transparency and auditability. This dual-layer model is valuable in government services, healthcare systems, and enterprise collaborations, where accountability must coexist with confidentiality.

Table 7.1 provides a detailed comparison of blockchain types.

Blockchain Types Comparison

Characteristic	Public	Private	Consortium	Hybrid
Decentralization	Fully decentralized	Centralized	Semi-decentralized	Combination
Consensus	PoW, PoS, DPoS	PBFT, RAFT, Custom	PBFT, PoA, RAFT	Flexible
Governance	Community-driven	Central authority	Shared governance	Mixed
Performance	Low to moderate	High	Moderate to high	Moderate
Use Cases	Cryptocurrencies, DeFi, NFTs	Supply chain, healthcare	Inter-bank, trade finance	Government, healthcare

Fig. 7.3 Comparison of blockchain types: public, private, consortium, and hybrid

7.1.4 Consensus Mechanisms

Consensus mechanisms form the backbone of blockchain systems by ensuring agreement on the state of the ledger across distributed nodes, even in adversarial environments. Different mechanisms balance decentralisation, security, scalability, and energy efficiency in distinct ways [5]:

Proof of Work (PoW): PoW requires participants (miners) to solve computationally intensive cryptographic puzzles to add new blocks. Bitcoin pioneered PoW, offering strong security guarantees by making attacks economically unfeasible: an attacker would need to control over 51% of the network's total hashing power. However, the method is highly resource-intensive, consuming enormous amounts of energy and limiting throughput to a few transactions per second.

Proof of Stake (PoS): The primary security mechanism in proof of stake is to replace energy expenditure with an economic stake. In PoS, validators are chosen to prescribe and validate blocks based on the tokens and commit as collateral. Penalties such as the slashing of their staked assets are assigned in case any malicious behaviour is detected. In this way, PoS creates network security while reducing energy needs.

Delegated Proof of Stake (DPoS): DPoS enables token holders to vote and select delegates and witnesses who are responsible for approving blockchain transactions. With comparison to PoW or PoS, DPoS obtains better throughput and lower latency.

Figure 7.4 demonstrates various consensus mechanisms that secure blockchain networks, including PoW, PoS, DPoS, PBFT, PoA, and Hybrid/DAG.

Practical Byzantine Fault Tolerance (PBFT): PBFT is designed to work on a blockchain network where the end users are already known and authenticated. Even when some of the nodes in a blockchain network fail or behave maliciously, PBFT ensures that nodes can agree on a block while maintaining high throughput. The

Table 7.1 Comparison of blockchain types

Blockchain type	Decentralisation	Consensus mechanisms commonly used	Governance	Performance (speed/ scalability)	Typical use cases
Public	Fully decentralised, open to all participants	Proof of Work (PoW), Proof of Stake (PoS), Delegated Proof of Stake (DPoS)	Community-driven, protocol-based	Low to moderate (limited throughput, higher latency)	Cryptocurrencies (Bitcoin), decentralised finance (Ethereum), NFTs
Private	Centralised or controlled by a single organisation	Practical Byzantine Fault Tolerance (PBFT), RAFT, custom consensus	Central authority or owner	High (fast transactions, scalable for enterprise)	Supply chain management, healthcare records, enterprise data sharing
Consortium	Semi-decentralised, controlled by group of organisations	PBFT, Proof of Authority (PoA), RAFT	Shared governance among members	Moderate to high (depends on number of participants)	Inter-bank settlements (R3 Corda), trade finance, cross-institution collaboration
Hybrid	Combination of public and private	Flexible (can integrate PoW/ PoS for public layer + PBFT/ PoA for private layer)	Mixed: internal governance with selective public oversight	Moderate (balance between transparency and efficiency)	Government services, healthcare, cross-border digital identity, regulated industries

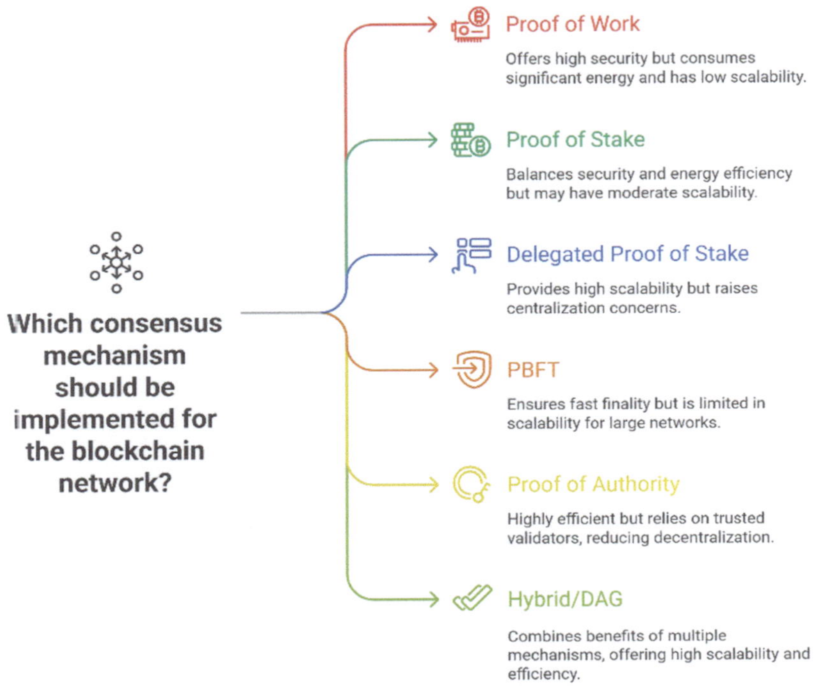

Fig. 7.4 Consensus mechanisms for blockchain networks

issue with PBFT lies in its scalability, as it can only support up to hundreds of nodes, but not a large number of nodes, which is often the case in real-world scenarios.

Table 7.2 provides a comparison between consensus mechanisms.

7.1.5 Evolution of Blockchain

The evolution of blockchain technology can be broadly categorised into three distinct generations. The subsequent generation addresses the shortcomings of the previous generations and enhances its quality [6].

Blockchain 1.0: The first generation, Blockchain 1.0, dates back to the cryptocurrency concepts and applications, such as Bitcoin. Cryptocurrencies operate based on the peer-to-peer transfer of value, eliminating the need for intermediaries.

Blockchain 2.0: Blockchain 2.0 focuses on smart contracts, where automated agreements and programmable money come into play. Ethereum provides a platform for decentralised applications to implement complex business logic.

Table 7.2 Comparison of consensus mechanisms

Consensus mechanism	Energy efficiency	Scalability	Security	Decentralisation	Example platforms
Proof of Work (PoW)	Very Low	Low (7–15 TPS)	Very High (economic cost of attack)	High	Bitcoin
Proof of Stake (PoS)	High	Moderate–high	High (stake slashing)	High	Ethereum 2.0, Cardano
Delegated PoS (DPoS)	High	High (thousands TPS)	Moderate (delegate collusion risk)	Moderate	EOS, TRON
PBFT	High	Moderate (hundreds nodes)	High (Byzantine fault tolerance)	Low (permissioned)	Hyperledger Fabric
Proof of Authority (PoA)	Very High	High	Moderate (validator trust)	Low (centralised validators)	VeChain, Microsoft Azure Blockchain
Hybrid/DAG	High	Very high (parallel validation)	Varies by design	Varies	Decred (PoW + PoS), IOTA (DAG)

Blockchain 3.0: This variation of blockchain is dedicated to enhancing the sustainability, availability, and interoperability of existing blockchain systems, so that they can be better equipped for mainstream use. There are many blockchain techniques and platforms, such as Cardano, Polkadot, and Solana, that have implemented advanced consensus mechanisms and governance protocols.

Layer-2 Solutions address scalability through additional functionality built on existing networks. Payment channels enable off-chain transactions periodically settled on main blockchains, dramatically increasing throughput.

Table 7.3 summarises the feature-wise differences among the blockchain technologies.

Figure 7.5 presents the technological evolution of blockchain.

7.1.6 Limitations and Challenges

The major issue with blockchain lies in its scalability constraints. This is the most significant issue that the technology industry needs to overcome. The blockchain systems can generally optimise the maximum of the three key properties: security, scalability, and decentralisation.

Table 7.3 Evolution of blockchain technologies

Generation	Key features	Limitations	Example platforms	Use cases
Blockchain 1.0	Peer-to-peer value transfer, decentralised digital currency	Limited scalability, no programmability, high energy use	Bitcoin	Digital payments, store of value
Blockchain 2.0	Smart contracts, decentralised applications (dApps), programmable money	Network congestion, high gas fees, limited interoperability	Ethereum, NEO	DeFi, tokenization, automated agreements
Blockchain 3.0	Scalability, interoperability, governance models, eco-friendly consensus	Complexity of protocols, still maturing ecosystems	Cardano, Polkadot, Solana	Cross-chain applications, enterprise adoption, sustainable dApps
Layer-2 Solutions	Off-chain scaling (payment channels, rollups), faster transactions, reduced costs	Dependence on underlying Layer-1 security, added complexity	Lightning Network (Bitcoin), Optimism, Arbitrum	Micropayments, high-frequency trading, gaming, NFT scalability

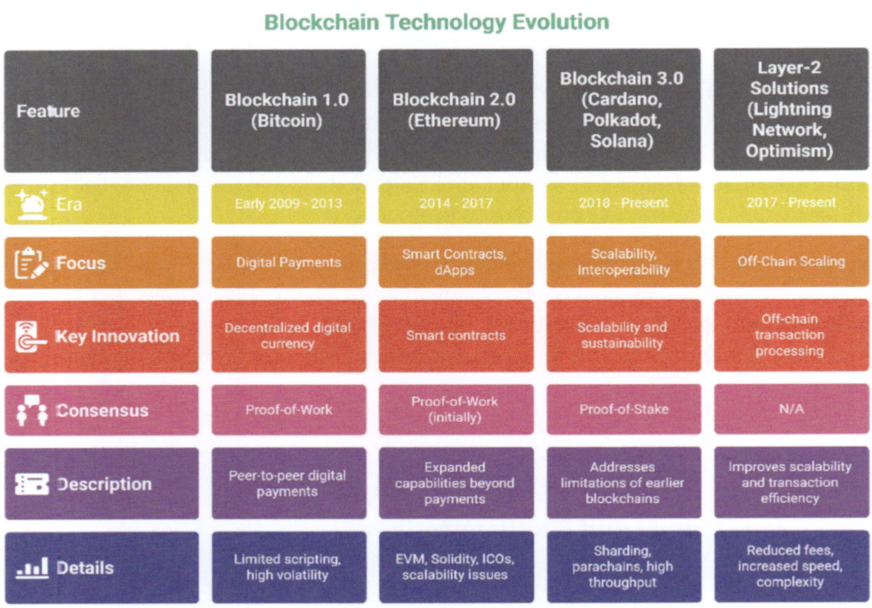

Fig. 7.5 Evolution of blockchain technology: from Blockchain 1.0 (Bitcoin) to Blockchain 3.0 and Layer-2 solutions

The next issue with blockchain is energy consumption. Due to the high computational resource requirements, energy consumption affects proof-of-work networks. Another issue with blockchain lies in the user experience barrier. This occurs due to the complexity of key management and the use of technical terminology. Another significantly increasing issue with blockchain technologies is the uncertainty in traditional regulatory organisations, as technological innovation in this domain is advancing at a higher rate than the legal frameworks.

Energy consumption concerns primarily affect proof-of-work networks, which require substantial computational resources. User experience barriers limit adoption through complex key management and technical terminology. Regulatory uncertainty creates challenges as legal frameworks struggle to keep pace with technological innovation.

7.2 Cryptocurrencies

7.2.1 Introduction

Cryptocurrencies are the first and most successful application of blockchain technology. It has almost become synonymous with the blockchain term. Cryptocurrencies utilise cryptographic algorithms to validate transactions in a distributed manner, control and create new data units, and provide security in transactions [7].

Cryptocurrencies are a relatively new concept in the financial monitoring system. The traditional banking system, backed by the government, depends on its currency value being determined by the fiat currency factors. Cryptocurrency derives its value from the user-connected network, market dynamics, and the utility of that particular cryptocurrency. It consists of store-of-value assets like Bitcoin and programmable platforms like Ethereum. Cryptocurrencies are a growing monetary system that has evolved from zero in 2009 to a market value of over $1 trillion now.

7.2.2 Operation of Cryptocurrencies

The operation of cryptocurrencies is based on the concept of the centralised currency system. In this enterprise, the current system values a currency based on three main factors: planning coin creation, action validation, and key management [8].

7.2.2.1 Creation and Mining

The concept of cryptography is diverse. For example, proof-of-work systems, such as Bitcoin, create new coins of currency through the mining process. Similarly, proof-of-stake systems, such as Ethereum 2.0, create new cryptocurrency tokens through a validation process. In this way, Bitcoin rewards the provider of computational resources, and Ethereum 2.0 rewards participants for staking existing tokens.

Mining operations are a costly endeavour, as they require new hardware for competing and solving cryptographic algorithms to gain new blocks. In this competition, full minors are provided with the benefits of new cryptocurrency, along with the transaction fees.

7.2.2.2 Transaction Process

The transaction process comes into play when we need to transfer digital messages specifying cryptocurrency transfers between different users and their corresponding addresses. A transaction takes input from previous transactions and outputs information to the recipient, including the amount and digital signatures for sender authorisation.

Valid transactions are distributed through the blockchain network and reach peer-to-peer connections. The consensus mechanism also works and shapes this process by confirming transactions and putting them into blocks.

7.2.2.3 Wallets and Keys

The wallets are a storage for keeping the private key in the context of cryptocurrency. There are two kinds of wallets: hot wallets and cold wallets. A hot wallet connected to the Internet provides availability, but it is prone to online attacks. Cold wallets are always offline to maximise security.

Key management is one of the most challenging aspects of cryptocurrency adoption. Lost private keys result in permanently inaccessible funds, while compromised keys enable irreversible unauthorised transactions. Hierarchical Deterministic (HD) wallets generate multiple addresses from a single seed phrase, simplifying backup while maintaining security.

7.2.3 Major Cryptocurrencies

The cryptocurrency domain consists of a diverse range of assets [9].

Table 7.4 Operation of cryptocurrencies and major types

Aspect	Key points	Examples
Creation and mining	PoW: Mining with computational puzzles; rewards in new coins + fees. PoS: Validation via staking tokens; rewards proportional to stake. ASIC hardware improves efficiency but risks centralization	Bitcoin (PoW), Ethereum 2.0 (PoS)
Transaction process	Transactions include inputs, outputs, digital signatures. Broadcast through P2P networks. Confirmed via consensus into blocks; more confirmations = stronger finality	Bitcoin, Ethereum
Wallets and keys	Wallets manage private keys, not coins directly. Hot wallets = convenient but online risk. Cold wallets = offline, highly secure. HD wallets derive multiple addresses from one seed	MetaMask (hot), Ledger Nano (cold)
Major cryptocurrencies	**Bitcoin**—Decentralised store of value, 21 M supply cap. **Ethereum**—Smart contracts, dApps, EVM ecosystem. **Stablecoins**—Fiat-pegged (e.g. USDT, USDC). **CBDCs**—State-issued digital currencies (e.g. Digital Yuan)	BTC, ETH, USDT, USDC, Digital Yuan

Bitcoin was the first cryptocurrency to gain global popularity since its release in 2009. The over 2 million Bitcoin supply has made a significant contribution to cryptocurrency usage and distribution. A summary of the operation of cryptocurrencies and major types is provided in Table 7.4.

Ethereum introduced programmable blockchain capabilities through smart contracts, enabling decentralised application development spanning finance, gaming, and digital art. EVM compatibility has created vast ecosystems of compatible networks and development tools.

Stablecoins like Tether (USDT) and USD Coin (USDC) maintain approximate parity with fiat currencies through various mechanisms, serving as bridges between traditional finance and cryptocurrency ecosystems.

Central Bank Digital Currencies (CBDCs) represent government-issued digital currencies combining cryptocurrency benefits with traditional monetary policy control, with China's digital yuan leading global pilot programs.

7.2.4 Risks and Challenges

Price volatility represents significant adoption challenges, with daily movements of 10–20% common for major cryptocurrencies. Security risks encompass exchange hacks, smart contract vulnerabilities, and phishing attacks targeting user credentials.

When demand is high for a certain period, scalability becomes the most crucial consideration for cryptocurrencies [10].

7.2.5 *Regulation and Global Perspectives*

Different countries have varying regulations. In the United States, we can see different states or regions with overlapping rules implemented by their governing bodies. However, the European Union has established framework rules and regulations to be implemented by all countries under its jurisdiction. In Asia, we observe various approaches to cryptocurrencies. For example, China has imposed a comprehensive ban, and Japan has provided a detailed, documented legal set-up for cryptocurrencies.

7.2.6 *Future of Cryptocurrencies*

The future of cryptocurrency is a challenging domain to predict, as it is highly volatile and heavily dependent on market shares; many traditional financial firms don't accept many of the existing cryptocurrencies. The mainstream and wide acceptance of cryptocurrency would occur when its price stabilises, the scalability issues can be solved, and the user experience is enhanced. The layer to scaling is progressing for the new blockchain architecture. We also have to integrate traditional finance sectors with cryptocurrency trading and exchange factors.

7.3 Applications and Challenges of Blockchain

7.3.1 *Applications*

The application of blockchain has penetrated various domains of work, ranging from finance to healthcare, as well as supply chain management and governance [11–14].

The financial organisations have traditionally used blockchain for international payment and trading systems. Traditional financial services can utilise blockchain for decentralised finance management through smart contracts and trading, eliminating the need for third-party involvement.

Supply chain management is a technique which manages door-to-door delivery- or producer-to-consumer-based management. Modern supply chain management firms utilise blockchain to provide end-to-end management of their products, ensuring a required level of transparency. For example, Walmart utilises blockchain technology to track the delivery from warehouses to retail stores.

Similarly, the healthcare domain utilises blockchain for pharmaceutical supply chain management, spanning from factories to medical shops, as well as for managing medical records. Blockchain can also be used to manage health records, allowing patients to have a certain level of autonomy over their data while maintaining privacy.

Blockchain is also used in digital identity and governance, where an individual's identity is controlled by digital identities and is not authorised by any centralised entity. For example, Estonia's residency program enables digital governance which goes beyond geographic boundaries using blockchain technology.

7.3.2 Challenges

Every technology faces crises and challenges. Blockchain also faces numerous challenges regarding sustainability, interoperability, scalability, and governance [15].

Scalability Challenges: Scalability challenges persist for blockchain technology; all new technologies are being experimented with to address the challenge of scalability. In most cases, improving one parameter, such as speed, often reduces the efficiency of other parameters, such as security or decentralisation.

Interoperability Limits: Generally, different blockchain systems don't operate seamlessly among themselves. This has created an issue with making a unified network of blockchains.

Governance Issues: Blockchains are decentralised, lacking centralised governance. Although this has been an advantage, it also presents challenges in managing upgrades and resolving conflicts.

Legal and Regulatory Rules: As blockchains are at the source of decentralised finance, the legal and regulatory bodies of different governments have different laws to manage them. These laws usually change across different countries and regions.

Energy Use: Cryptocurrencies require a robust and consistent computational system that consumes significant power over time.

7.3.3 Future of Blockchain Adoption

The future of blockchain adoption lies in its integration with existing enterprise systems. Regulations adopted by various authoritarian frameworks are maturing to address concerns related to consumer protection and financial stability [16]. The convergence of technology with AI, IoT, and other emerging technologies will create new application possibilities.

7.4 Ethical, Security, and Sustainability Considerations

The blockchain technology and cryptocurrencies are entering a mature phase, and their real-world implications in the fields of security, ethics, and sustainability are being evaluated, considering their economic and social impact [17].

7.4.1 Privacy and Anonymity

Blockchain systems offer complex privacy models that differ from anonymous cash and identifiable digital payments. Anonymisation makes identities delinked from addresses, but intelligent transaction analysis techniques can deanonymise users' identities exposed by machine learning techniques of pattern recognition. Privacy-enhancing technologies, such as zero-knowledge proofs, aim to provide stronger anonymity guarantees.

7.4.2 Ethical Use of Cryptocurrencies

Cryptocurrency enables users to access the financial market and engage in trading without the support of traditional, government-backed banking systems. This has made cryptocurrencies often much less stable than traditional monetary systems. The technical complexities can also provide very limited access to the users of cryptocurrencies.

7.4.3 Cybersecurity Risks

Cybersecurity risk lies in the exchange security. In this exchange, the security of the centralised exchange controls cryptocurrency, which is vulnerable to cyberattacks. Similarly, smart contacts use centralised tools and applications that can identify and manipulate programming errors. Similarly, private key best security wallets raise security concerns, as users must protect themselves; they are assets that rely on the user's own protection, rather than the traditional banking system.

7.4.4 Environmental Sustainability and Energy Use

The energy use of blockchain is quite high. Bitcoin alone consumes more electricity than a few small countries. The proof-of-work mining process in blockchain requires significant energy consumption, raising concerns about environmental sustainability. Although the proof-of-stake consensus mechanism reduces energy consumption, integrating renewable energy resources and utilising them should be considered for a sustainable environmental model.

7.4.5 Criminal Use Versus Regulation

There are reports of the misuse of cryptocurrency technology for ransomware, money laundering, and the wrongful evasion of sanctions. New regulations are being introduced to address these illicit activities and prevent their criminal use. Law enforcement authorities are now providing specialised training and blockchain analysis to improve their framework of crime analysis.

7.5 Emerging Trends and Research Frontiers

Blockchain research and development are giving way to new domains of work where blockchain technologies can have a significant impact. Financial infrastructure, digital ecosystems, governance models, and other related aspects [18].

7.5.1 Web 3.0

Web 3.0 provides the vision of a new Internet system where users will connect with each other with decentralised network connectivity. In Web 3.0, users control their digital assets, their data and identities without any centralised system control. In this context, blockchain can be used to decentralise traditional Internet services.

7.5.2 Decentralised Finance (DeFi)

DeFi finance is a concept that utilises blockchain to decentralise the financial system, eliminating the need for traditional government-based banks. Smart contracts are used for borrowing, trading, and getting derivatives without any centralised support.

7.5.3 Non-Fungible Tokens (NFTs)

Non-Fungible Tokens (NFTs) are digital assets which are provided by blockchain standards like ERC-721. NFTs have grown steadily and at a rapid rate in the creative industries, as they provide monetisation channels for them. However, the problem with NFT lies with the market volatility and copyright disputes.

7.5.4 Central Bank Digital Currencies (CBDCs)

Central Bank Digital Currencies (CBDCs) are issued by government-backed banks and have the same value as a physical currency. As they are backed by the central banks, they provide financial broad inclusion to the masses with improved policy and faster payments.

7.5.5 Cross-Chain Interoperability

Blockchain technology is typically not interoperable and often operates independently. There are cross-chain interoperability techniques, such as atomic swaps, cross-chain bridges, and side chains, which aim to interoperate different blockchains while removing the limitations and managing heterogeneous consensus mechanisms.

7.6 Case Studies of Blockchain Deployments

The practical impact of blockchain is best illustrated through real-world implementations across diverse sectors. Case studies highlight how organisations and governments leverage blockchain to enhance efficiency, transparency, and security in critical domains [19].

7.6.1 Financial Services: JPMorgan Chase's JPM Coin

JPMorgan developed Quorum and JPM Coin enabling blockchain-based instant transfers between institutional clients, showcasing how permissioned blockchains can coexist with regulatory frameworks while addressing privacy concerns.

7.6.2 *Supply Chain Management: IBM Food Trust*

Built on Hyperledger Fabric, IBM Food Trust provides immutable ledgers for tracking food products. Partners like Walmart use the system reducing contamination source identification time from weeks to seconds.

7.6.3 *Healthcare: Estonia's National eHealth System*

Estonia pioneered blockchain integration in national healthcare infrastructure, securing patient health records while ensuring data is tamper-proof and accessible only to authorised parties.

7.6.4 *Government: Dubai Blockchain Strategy*

Dubai's Blockchain Strategy 2020 aims to migrate all government documents to blockchain platforms for property registration, visa issuance, and business licensing, demonstrating operational efficiency improvements.

7.7 Conclusion and Future Outlook

Blockchain and cryptocurrencies represent a paradigm shift in digital trust, value exchange, and governance. While DeFi, NFTs, CBDCs, and interoperability continue reshaping global systems, challenges of scalability, energy consumption, regulation, and ethical use remain critical.

The next decade will likely see blockchain convergence with AI, edge computing, and quantum technologies, leading to novel architectures and applications. Research is expected to intensify in green consensus mechanisms, privacy-preserving techniques, and cross-border legal harmonisation.

Blockchain is transitioning from disruptive innovation to enabling infrastructure—its ultimate impact will depend on balancing decentralisation, efficiency, regulation, and sustainability.

7.8 Summary

This chapter provides a detailed exploration of blockchain and cryptocurrencies, beginning with the foundational principles of blockchain such as its definition, functioning, types, consensus mechanisms like proof of work and proof of stake, and the role of smart contracts, while also tracing its evolution and limitations. It then explains the operation of cryptocurrencies, covering mining, transaction processes, wallets and keys, major digital currencies like Bitcoin and Ethereum, as well as associated risks, regulations, and adoption challenges. The chapter further examines blockchain applications in finance, supply chain, healthcare, and governance, alongside issues of scalability, interoperability, and regulation. Ethical, security, and sustainability considerations are addressed, including privacy, anonymity, cybersecurity risks, environmental concerns, and the tension between criminal use and regulation. Emerging trends such as Web3, decentralised finance (DeFi), non-fungible tokens (NFTs), central bank digital currencies (CBDCs), and cross-chain interoperability are highlighted as drivers of digital transformation. Finally, real-world case studies—including JPMorgan Chase's JPM Coin, IBM Food Trust, Estonia's blockchain-enabled eHealth system, and Dubai's blockchain strategy—demonstrate practical deployments. Overall, the chapter integrates theoretical foundations, practical implementations, and future directions, presenting blockchain and cryptocurrencies as both a disruptive force and a subject of ongoing research and debate.

References

1. Kumar C, Kumar KA, Gupta S, Sardar TH. Cyberbullying detection based on the fusion of DistilBERT and SIMHASH Technique. In: 2024 2nd international conference on artificial intelligence and machine learning applications theme: healthcare and internet of things (AIMLA). IEEE; 2024. p. 1–4.
2. Ramachandran M. Introduction to blockchain concepts. In: Blockchain engineering: secure, sustainable frameworks for healthcare applications, vol. 18. Singapore: Springer Nature Singapore; 2025. p. 3–33.
3. Martinez D, Magdalena L, Savitri AN. AI and blockchain integration: enhancing security and transparency in financial transactions. Int Trans Artif Intell. 2024;3(1):11–20.
4. Mlika F, Karoui W, Romdhane LB. Blockchain solutions for trustworthy decentralization in social networks. Comput Netw. 2024;1(244): 110336.
5. Kumbhar V, Keskar V, Avhankar MS, Yarde P, Walhhekar R, Kumbhar SA. Reconnoitering blockchain consensus mechanisms. In: Fostering machine learning and IoT for blockchain technology: smart cities applications, vol. 1. Singapore: Springer Nature Singapore; 2025. p. 65–114.
6. Bandarupalli G. The evolution of blockchain security and examining machine learning's impact on ethereum fraud detection. In: 2025 17th International conference on electronics, computers and artificial intelligence (ECAI). IEEE; 2025. p. 1–6.
7. Budish E. Trust at scale: the economic limits of cryptocurrencies and blockchains. Q J Econ. 2025;140(1):1–62.

8. Bhullar PS, Joshi M, Sharma S, Phan DH, Nguyen AL. Global taxation issues in cryptocurrencies–a synthesis of literature and future research agenda. J Account Lit. 2025.
9. Bhattacherjee P, Mishra S, Kang SH. Extreme frequency connectedness, determinants and portfolio analysis of major cryptocurrencies: insights from quantile time-frequency approach. Q Rev Econ Finance. 2025;1(100): 101974.
10. Magalhães J. Current vulnerabilities and risks of blockchain: a systematic literature review. In: Digital technologies and transformation in business, industry and organizations; 2025, p. 159–75.
11. Igonor OS, Amin MB, Garg S. The application of blockchain technology in the field of digital forensics: a literature review. Blockchains. 2025;3(1):5.
12. Felemban S, Yiu TW, Harvey M, Lim H, Zhu J, Bolton D, Ischia J. Current application of blockchain technology in healthcare and its potential roles in Urology. BJU Int. 2025.
13. Mohapatra S, Sainath B, KC A, Lal H, K NR, Bhandari G, Nyika J, R S. Application of blockchain technology in the agri-food system: a systematic bibliometric visualization analysis and policy imperatives. J Agribus Dev Emerg Econ. 2025;15(2):288–310.
14. Li Z, Wei J. Policy optimization and practical applications in blockchain-enabled electric vehicle battery recycling system. Comput Ind Eng. 2025;1(204): 111119.
15. Bendig D, Charlet M. Opportunities and challenges of blockchain for multi-sided platforms. Electron Mark. 2025;35(1):25.
16. Lumineau F, Shang G, Swaminathan JM, Tsoukalas G, Wagner SM, Zhao JL. Charting the future of blockchain in operations and supply chain management: opportunities and challenges. J Oper Manag. 2025.
17. Mustafa G, Rafiq W, Jhamat N, Arshad Z, Rana FA. Blockchain-based governance models in e-government: a comprehensive framework for legal, technical, ethical and security considerations. Int J Law Manag. 2025;67(1):37–55.
18. Zhang J, Gao Y, Alam S. A visual analysis of hotspots and frontiers in blockchain based on scientific knowledge Map. J Knowl Econ. 2025;8:1–38.
19. Dwivedi B, Newar N, Mahajan A, Jimmy AS. Case studies of blockchain and AI integration in medical systems. In: AI and blockchain applications for privacy and security in smart medical systems. IGI Global Scientific Publishing; 2025. p. 133–160.

Chapter 8
Cybersecurity and Ethical Hacking

8.1 Fundamentals of Cybersecurity

8.1.1 Basic Terminology and Core Concepts

The cybersecurity domain consists of a set of vocabulary or terms which define its core concepts. To understand the ecosystem of cybersecurity, it is essential to clearly understand these terms and their meanings [1].

Threat: A threat is any event that has the capacity to negatively impact organisational operations or damage its assets and individuals through unauthorised access or denial of service. Here, threat generally refers to human errors or redesigning automated security destruction systems created by individuals, a group, or any organisation.

Vulnerabilities: Vulnerabilities refer to the point of weakness in a cyber system. This weakness may be due to a design error, an issue in software code, processes or physical infrastructure. A threat can only occur if one or more vulnerabilities exist.

Risk: A risk is the potential loss or damage that can occur due to the successful execution of a threat through one or more of its vulnerabilities. Risk analysis is a theoretical assessment of the likelihood of potential attacks and their subsequent damage or loss.

Defence in depth: Defence in depth is a concept where the cyber system is secured using multiple layers of security controls. This idea of Defence in depth is based on the philosophy that a single security layer can never be infallible. Therefore, we require a multi-layer protection system.

T. H. Sardar, *The Spectrum of Computer Science: Emerging Technologies and Trends*,
Studies in Smart Technologies, https://doi.org/10.1007/978-3-032-19632-3_8

8.1.2 The CIA Triad: Pillars of Information Security

The CIA triad is the pillar of information security, on which the practices and decision-making of cybersecurity are built. Security requirement evaluation, protective measure designing, and evaluation and judgment of security impacts can be developed based on the CIA triad framework [2]. Now, let us explain the components of CIA.

Confidentiality of CIA refers to the fact that data is only accessible by a process from authorised entity(ies) or individual(s). Services such as encryption, stenography, access control mechanisms, and physical security systems fall under this category. Maintaining confidentiality is challenging due to the rapid generation of data and the complexity of internet-based data sharing, as well as the intricate and ever-evolving design of malicious threats and attacks.

Integrity refers to the trustworthiness of the data in its lifecycle (from its creation to its deletion). Data integrity can be maintained if the data is not corrupted or deleted by an unauthorised source. Widely used services employed for maintaining integrity are digital signatures, cryptographic checksums, and change management processes. Maintaining Data integrity and system integrity is challenging due to sophisticated attack mechanisms, such as insider threats, advanced persistent threats, and supply chain compromisers, which have emerged in the cybersecurity landscape.

Availability refers to the system when the data are always accessible and usable for an authorised user. This principle addresses disruptions to normal operations, whether caused by system failures, natural disasters, or deliberate attacks. Availability controls include redundancy, fault tolerance, backup and recovery systems, and incident response capabilities.

The availability imperative has grown more critical as organisations become increasingly dependent on digital systems for core business functions. Distributed denial-of-service (DDoS) attacks, ransomware, and infrastructure failures can have immediate and severe business impacts. Modern availability strategies must address both traditional high-availability requirements and emerging challenges such as cloud service dependencies and global supply chain vulnerabilities. The CIA Triad is presented using Fig. 8.1.

8.1.3 Cybersecurity Frameworks and Standards

The cybersecurity field has matured through the development of comprehensive frameworks and standards that provide structured approaches to security management, risk assessment, and compliance. These frameworks serve as blueprints for organisations seeking to establish or improve their cybersecurity postures [3].

The most prominent standard for cybersecurity is provided by the National Institute of Standards and Technology (NIST). This model was developed following a detailed discussion with government officials and industry professionals. NIST

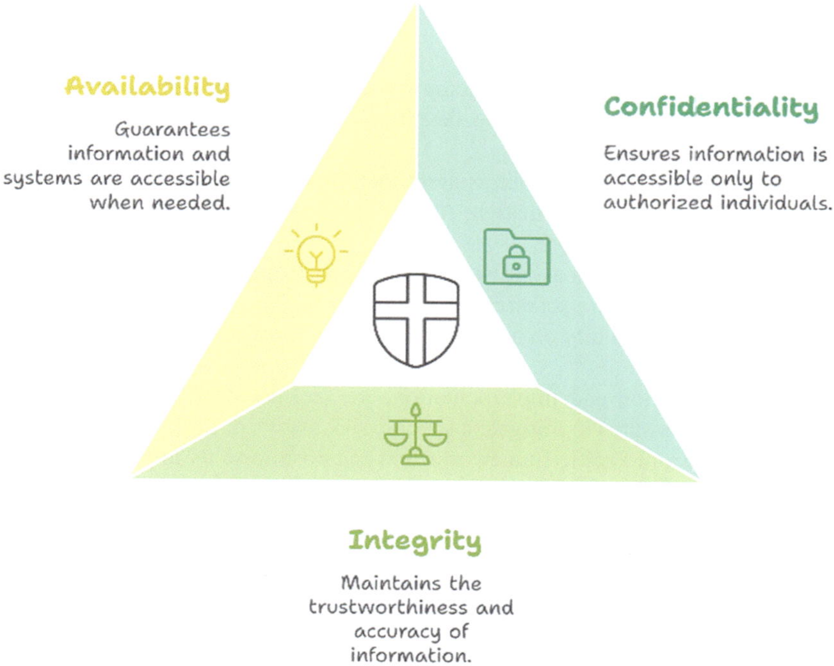

Fig. 8.1 the CIA triad

has divided all cybersecurity and its related processes into 5 specific procedures. Procedures are Identify, Protect, Detect, Respond, and Recover.

Using NIST, we can develop industry-specific requirements, business scenarios, and context-based cybersecurity solutions. This Framework has gained its popularity and has been adopted globally due to its technology-neutral and outcomes-specific approach.

Information Security Management Systems (ISMS) have been presented by ISO 270001. ISMS has created a framework that enables organisations to establish, develop, deploy, and maintain an information security management system. ISO 270001 has a multiple-step process that an organisation follows to implement security controls based on their evaluation of certain security risks.

There are other frameworks, such as GDPR, HIPAA, and PCI DSS, which are developed for certain industries. GDPR, for example, has a global presence to protect personal data and measures for European Union citizens.

8.1.4 Cybersecurity Lifecycle

The cybersecurity lifestyle is essentially a set of processes that must be used to evaluate and improve the life cycle development of cybersecurity systems in an organisational context [4].

The first phase of this lifestyle is asset identification and classification. In this phase, we identify the information Assets and required processes which need protection from attacks.

The next phase is the risk assessment. First, we try to understand the probability of a potential threat and its impact on cybersecurity assets. This assessment evaluates the impact of the rate by quality on quantitative models best suited to the organisation's size.

In control selection and implementation, we seek to identify suitable security techniques and measures to mitigate a specific risk assessed. Essentially, we adopt a specific control from Pacific frameworks and can customise its implementation for a scenario-based evaluation.

The monitoring and detection phase provides security information and event management, planning for vulnerability, integration of threat intelligence, and behavioural analytics. Today's monitoring systems utilise machine learning and artificial intelligence to recognise anomalous patterns in our data and identify them as potential threats.

Figure 8.2 shows a circular lifecycle diagram of cybersecurity stages.

Fig. 8.2 A circular lifecycle diagram of cybersecurity stages

8.2 Threats and Vulnerabilities

8.2.1 *Human and Social Engineering Threats*

The biggest reason, to date, for cyber assets is human activity. Humans can twist existing technology and misuse it, employ psychological tricks, and manipulate organisational dynamics. Humans can gain unauthorised access to an information system through social engineering attacks by exploiting trust relationships, cognitive, and information asymmetries [5].

Phishing attacks continue to be one of the most popular and effective forms of social engineering. Fraudulent communications are usually used in these attacks to fool recipients into disclosing private information or taking actions that jeopardise security. Sophisticated multi-channel attacks, involving voice calls (vishing), text messages (smishing), and social media platforms, are now incorporated into modern phishing campaigns, which have evolved beyond basic email-based tactics.

Spear phishing is an advanced phishing technique that uses individually tailored information obtained through reconnaissance to target specific individuals or organisations. One particularly harmful type that targets financial processes and frequently results in large financial losses is business email compromise (BEC). Phishing attacks are successful because they exploit human psychological traits, such as reciprocity, urgency, and authority bias.

To build trust and elicit information from targets, pretexting involves creating scenarios. To persuade targets to divulge private information or allow unauthorised access, attackers may pose as trustworthy individuals, fabricate emergencies, or create elaborate backstories. Pretexting attacks are particularly challenging to identify and stop because they often require extensive planning and research.

Insider threats are security risks that originate from individuals who have been granted permission to access organisational resources. These threats could be malicious (employees purposefully harming others), negligent (employees carelessly causing security incidents), or compromised (legitimate users whose credentials have been stolen or coerced). Sophisticated behavioural analytics, privilege monitoring, and cultural approaches that strike a balance between security and employee privacy and trust are necessary due to the complexity of insider threat detection.

8.2.2 *Malware and Ransomware*

Many software programs have been developed to create a wide range of attacks, posing a threat to information systems [6]. From straightforward proof-of-concept applications to complex, for-profit attack platforms that can elude sophisticated detection systems, the malware landscape has changed over time. Viruses (self-replicating code that attaches to other programs), worms (self-propagating malware that spreads across networks), and trojans (malware that poses as trustworthy software) are examples of traditional malware categories. Although these classifications are still applicable, contemporary malware often combines multiple methods and exhibits polymorphic or metamorphic traits, enabling it to evade detection systems that rely on signatures.

Advanced persistent threats (APTs) are complex, multi-phase attacks typically attributed to advanced criminal organisations or nation-state actors. APTs are distinguished by their ability to target high-value data or systems with precision, persistence, and stealth. Custom malware, zero-day exploits, and meticulous operational security measures are frequently employed in these attacks to maintain long-term access to compromised environments.

One of the biggest cybersecurity risks nowadays is ransomware, which combines advanced technology with immediate financial consequences. Modern ransomware attacks often follow a double-extortion model, encrypting victim data while simultaneously exfiltrating sensitive information to increase pressure for payment. Ransomware-as-a-service models have provided a less expensive, cloud-based platform that can be rented, even by small organisations.

8.2.3 System and Network Vulnerabilities

A large number of cyberattacks exploit weaknesses within hardware and software systems [7]. Such weaknesses may arise from poor design, implementation errors, incorrect configuration settings, or the inherent complexity of today's computer environments.

Among these, software vulnerabilities form a major category, covering both design flaws and coding errors that attackers can leverage to gain unauthorised access or disrupt normal system operations. Common vulnerability categories include buffer overflows, injection flaws, cross-site scripting, and insecure direct object references. The software vulnerability landscape is complicated by the use of third-party libraries, open-source components, and complex software supply chains that may introduce vulnerabilities beyond direct organisational control.

Network infrastructure vulnerabilities arise from misconfigurations, protocol weaknesses, and architectural design flaws. Network segmentation failures can enable lateral movement by attackers, while inadequate access controls may provide unauthorised network access. New attack surfaces and vulnerability categories have been brought about by the complexity of contemporary network architectures, such as software-defined networking and hybrid cloud environments.

Hardware vulnerabilities have gained increased attention following discoveries such as Spectre and Meltdown, which demonstrated fundamental flaws in processor design that could be exploited to bypass security boundaries. Because hardware flaws can impact millions of devices and are difficult to fix with software patches alone, they present unique challenges.

8.2.4 Emerging Threats

As technology advances, attack economics change, and geopolitical dynamics shift, the cybersecurity threat landscape keeps evolving [8]. Emerging threats often exploit the convergence of multiple technologies or target newly introduced systems and processes.

Artificial intelligence-powered attacks represent a significant evolution in threat capabilities, leveraging machine learning and automation to enhance attack effectiveness and scale. AI technologies enable more sophisticated social engineering through deepfake voice and video generation, automated vulnerability discovery, and adaptive attack strategies that can evade traditional detection methods.

Internet of Things (IoT) and cloud vulnerabilities have expanded attack surfaces far beyond traditional computing environments. IoT devices often lack robust security controls, creating entry points for network access and botnet recruitment. Cloud security challenges include misconfigurations, inadequate access controls, and shared responsibility model confusion.

Quantum computing, as it has been hypothetically claimed to date, can potentially compromise the cybersecurity frameworks of existing cryptographic systems. A hypothetical quantum computer, which may become available within a few decades, can decipher cryptographic algorithms, creating opportunities for future engineers to develop more secure security architectures and protocols.

8.3 Ethical Hacking

8.3.1 Overview and Legal Boundaries

Ethical hacking, also known as white-hat hacking, is the authorised practice of assessing systems and networks to identify security weaknesses before malicious attackers can exploit them [9]. By conducting controlled and approved tests, ethical hackers enhance an organisation's security posture, combining technical expertise with professional ethics and legal compliance.

The field emerged from the need to understand an attacker's perspective and tactics in order to build stronger defences. By simulating real-world attacks in a careful and responsible manner, ethical hackers provide valuable insights into system vulnerabilities, the effectiveness of existing protections, and the readiness of incident response strategies.

Legal frameworks surrounding ethical hacking vary widely across countries; however, in most cases, explicit permission is required before conducting any form of testing. For example, in the United States, the Computer Fraud and Abuse Act (CFAA) prohibit unauthorised access, and thus, written consent is required to conduct penetration testing. Ethical hackers also encounter complex issues in the world, such as the varying definitions of authorised security research and limitations on the export of security tools. The practice in this field is obligatory in following the scope of testing that is declared. This will provide assurance that operations are not conducted in a manner that may be considered illegal or contrary to the contract. Testing methods, timing, and data handling should also be planned carefully to minimise business impact and provide maximum benefit to security.

8.3.2 Types of Hackers

The diverse motives, levels of expertise, and moral orientations within the hacking community impact cybersecurity threat environments and defence strategies [10]. Understanding these differences can help organisations develop appropriate security controls and provide a context to threat modelling.

White-hat hackers do so with full consent and goodwill, focusing on improving security by finding vulnerabilities in a responsible manner and providing remediation recommendations. This category includes professional penetration testers, security researchers and bug bounty participants, who follow the responsible disclosure principle and collaborate with organisations to address security issues.

Black-hat hackers perform unauthorised activities out of a bad motive so as to gain a personal gain, cause harm or achieve other criminal objectives. Financial fraud, intellectual property theft, system destruction, and privacy violations are examples of black-hat activities. The motivations driving black-hat hacking include financial gain, political activism, personal grievances, and nation-state objectives.

Gray-hat hackers operate in ethical and legal ambiguity, sometimes discovering and disclosing vulnerabilities without explicit authorisation but without malicious intent. While gray-hat activities may ultimately benefit security, they raise legal and ethical concerns regarding unauthorised access and potential for unintended harm.

Figure 8.3 depicts a comparison infographic showing three hacker archetypes: White-hat (ethical hacker with laptop and shield icon), Black-hat (dark hoodie figure with red skull icon), Grey-hat (neutral tone with balance scales). Clean, educational vector style.

8.3.3 Penetration Testing Lifecycle

Professional penetration testing follows structured methodologies designed to ensure comprehensive, controlled, and repeatable assessments of security postures. These methodologies balance the need for realistic attack simulations with safety, legality, and business continuity requirements.

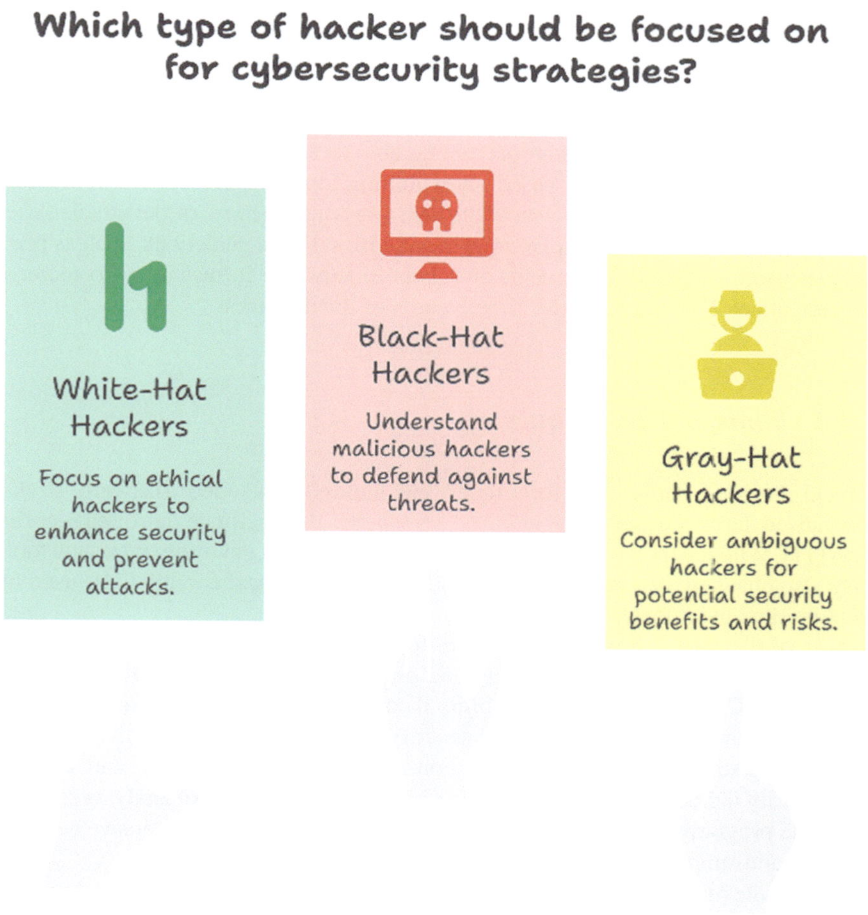

Fig. 8.3 A comparison infographic showing three hacker archetypes

Pre-engagement activities establish the foundation for successful penetration tests through scope definition, legal agreement execution, and stakeholder communication. This phase includes defining testing objectives, identifying target systems and networks, establishing rules of engagement, and coordinating with relevant organisational personnel to minimise business disruption.

Information gathering and reconnaissance involve collecting publicly available information about target organisations, systems, and personnel. Examples of this passive intelligence gathering include public database searches, social media reconnaissance and domain name system (DNS) enumeration. In the case of research and without contact with the actual targets, the objective will be to map the attack surface and identify possible entry points. Instead, vulnerability assessment and analysis is an active scanning and probing of systems to reveal vulnerabilities. The stage usually incorporates network scanning, port scanning, services scanning and automated vulnerability testing. The results are then checked and ranked to determine the most significant exploit attempts and be able to form an attack plan. Exploitation and post-exploitation are based on exploiting found vulnerabilities in order to obtain unauthorised access and show the possible consequences of the attack. After an exploit has worked, testers can experiment with subsequent actions, such as privilege escalation, lateral movement, or access to sensitive information, to recreate realistic attack pathways and the effectiveness of the defences.

8.3.4 *Tools and Techniques*

Ethical hacking tools and techniques encompass developed software packs, specialised hardware and well-organised approaches that facilitate thorough security tests [11]. With the increase in the complexity of computing environments, such tools have also changed, becoming more powerful but yet practical enough to be used by an individual in security.

One of the most complete penetration testing frameworks is Metasploit, which offers post-exploitation modules, payload generation, and automated exploitation capabilities. Comprehensive vulnerability databases, adaptable attack modules, and the ability to integrate with additional security tools are all features of the platform.

Understanding network communications, spotting security flaws, and looking into security incidents all depend on Wireshark's network protocol analysis capabilities The program offers strong filtering and analysis features for extensive network captures and supports hundreds of network protocols.

Burp Suite offers proxy, scanner, and manipulation tools to find web-based vulnerabilities with an emphasis on web application security testing. The platform offers a wide range of customisation options for intricate application testing scenarios and supports contemporary web technologies.

To manage the size and complexity of modern computing environments, modern penetration testing is depending more and more on automation, artificial intelligence, and cloud-based platforms.

8.3.5 *Bug Bounty Programs*

Bug bounty programs use a crowdsourced approach to uncover security vulnerabilities by drawing on the diverse skills and perspectives of security researchers worldwide [12]. These programs provide clear legal frameworks for conducting security research and reward participants with financial incentives for responsibly disclosing vulnerabilities.

The design of such programs can vary widely, with differences in their overall goals, the types of rewards offered, and the rules that define who can participate. While some programs cover entire organisational infrastructures, others concentrate on particular applications or services. Financial compensation, recognition, and professional growth opportunities are examples of reward systems; the amounts usually vary according to the degree of vulnerability and the difficulty of exploitation.

Bug bounty programs are ethically based on responsible disclosure practices, which mandate that researchers give affected organisations comprehensive vulnerability information prior to public disclosure. This approach balances the benefits of public security research with the need to protect organisations and users from active exploitation.

8.4 Security Mechanisms and Architectures

8.4.1 Cryptography and Encryption

Modern cybersecurity is mathematically based on cryptographic systems, which offer the technical means of achieving the goals of confidentiality, integrity, authentication, and non-repudiation [13]. The continuous arms race between security and attack capabilities is reflected in the development of cryptography from classical cyphers to quantum-resistant algorithms.

By using shared secret keys for encryption and decryption, symmetric cryptography makes it possible to process massive amounts of data quickly. The most widely used symmetric algorithm is Advanced Encryption Standard (AES), which provides strong security at manageable processing costs. Symmetric systems perform well in situations that call for high-throughput encryption, but they have trouble distributing keys in large-scale implementations.

Secure communications between parties without prior key sharing are made possible by asymmetric cryptography, which separates encryption and decryption capabilities using mathematically related key pairs. By resolving the key distribution issue that symmetric systems inevitably face, public-key systems such as elliptic curve cryptography (ECC), RSA, and others have transformed secure communications.

Hash functions play a central role in applications such as digital signatures and data integrity checks. Their operation consists of mathematical one-way transformations that convert any input to a fixed length digest. To be used in cryptography, though, such functions need to be strong enough to withstand usual attack techniques, including collision, preimage, second-preimage attacks.

Quantum computing is posing a threat to the underlying mathematics of existing public-key cryptography systems, since quantum algorithms may prove them unsound. The aim of post-quantum cryptography is thus to create new algorithms that are resistant to a quantum attack and yet as trustworthy and functional as traditional algorithms.

8.4.2 Authentication, Identity, and Access Control

Resource access control is based on the use of identity and access management (IAM) systems to make sure that the credentials used in the authentication process are part of the authorised user or system. Current IAM systems must be capable of managing the complexity of cloud computing services, mobile access, and distributed computing platforms without losing the balance between high levels of security and usability.

One strong aspect that can be employed is multi-factor authentication (MFA), wherein a user must submit two or more independent verification components, including something they know (a password), something they possess (a token), or something they are (biometrics). MFA causes attackers to trade off several factors simultaneously, making stolen or abused credentials less likely to occur.

RBAC also simplifies the security management further because it assigns permissions to the roles that are related to the job functions and not to individual users. This makes the process of granting permission to be consistent and administration more efficient. Zero-trust system architectures take the next step by abandoning the past concept of trusted locations of the network. With zero-trust, all access requests should be authenticated and authorised in real time, depending on the context and its risk.

8.4.3 Firewalls, IDS/IPS, and SIEM

Firewalls, intrusion detection and intrusion prevention systems (IDS/IPS) and Security Information and Event Management (SIEM) platforms are crucial to the prevention of malicious activity on network. Simple packet filtering has progressed to more sophisticated systems that can deep packet inspect, perform behavioural analysis, and also feed on threat intelligence feeds.

The next-generation firewall (NGFWs) incorporates traditional filtering and the intrusion prevention with the application-aware filtering and advanced threat protection. This allows to control traffic finely, recognising and controlling applications irrespective of port or protocol. IDS systems on the other hand monitors network and host activity to identify suspicious behaviour. Network based IDS are concerned with the traffic and content, whereas host based IDS investigate with system calls, file modifications, and user activity.

SIEM platforms also offer centralised visibility, which is by assembling and correlating various sources of security data. The advanced SIEMs use machine learning and threat intelligence as well as behavioural analytics to enhance the accuracy of detection as well as minimise the burden on security analysts.

8.4.4 Security by Design and Secure SDLC

Secure software development also focuses on the creation of protection at the systems level, as early as the design phase and not as an afterthought. This security by design strategy will be deployed throughout the software development life-cycle, such planning, design, deployment, and maintenance. This proactive approach is not only more effective, but more typically less expensive than attempting to add security to a system already constructed.

The threat modelling is a significant practice in first phases of the design because it ensures that future risks are identified and dealt with systematically. Models such as STRIDE (Spoofing, Tampering, Repudiation, Information Disclosure, Denial of Service, Elevation of Privilege) offer teams a way to reason about potential attack vectors and come up with suitable protections.

Lastly, secure coding practices include the use of rigorous programming practices and standards that are designed to avoid the most prevalent weaknesses in such that security is built directly into the software being created. Such practices are input validation, output encoding, parameterised queries, error handling, and secure session management. Automated static analysis tools can help identify potential security issues during development.

DevSecOps integrates security practices into DevOps workflows, emphasising automation, continuous integration, and security as code principles. This approach enables rapid, secure software delivery by incorporating security testing, compliance checking, and vulnerability scanning into automated deployment pipelines.

8.4.5 AI and ML in Cybersecurity

Artificial intelligence and machine learning technologies are revolutionising cybersecurity by providing automated threat detection, behavioural analysis, and adaptive defence mechanisms [14]. These technologies address the scale and complexity challenges facing modern cybersecurity operations while enabling new approaches to threat hunting and incident response.

Behavioural analytics leverage machine learning algorithms to establish baseline patterns of user and system behaviour, enabling the detection of anomalous activities that may indicate security incidents. These systems can identify subtle indicators of compromise that might evade traditional signature-based detection systems.

Threat intelligence platforms use natural language processing and machine learning to collect, process, and analyse threat information from diverse sources. AI-enhanced threat intelligence can identify emerging attack patterns, attribute activities to threat actors, and predict likely future attack vectors.

However, AI systems in cybersecurity also introduce new vulnerabilities and attack vectors. **Adversarial machine learning** attacks can manipulate AI systems through data poisoning, model inversion, and adversarial examples. Organisations must carefully consider the security implications of AI deployment while realiaing the benefits of enhanced security capabilities.

8.5 Case Studies

8.5.1 Major Data Breaches: Lessons from WannaCry, SolarWinds, and Equifax

Examining significant cybersecurity incidents provides invaluable insights into attack methodologies, organisational vulnerabilities, and the cascading effects of security failures [15]. These case studies demonstrate the evolution of cyber threats and the critical importance of comprehensive security strategies.

Figure 8.4 infographic showing three famous cyber attacks: WannaCry ransomware (red lock icon over hospital computer), SolarWinds supply chain attack (infected software update pipeline icon), Equifax data breach (database with leaking personal data). Modern infographic, professional educational style.

The WannaCry Ransomware Campaign (2017) represents one of the most significant global cybersecurity incidents, affecting hundreds of thousands of computers across more than 150 countries within days. The attack took advantage of a Windows vulnerability that was first identified by the NSA and then made public by the Shadow Brokers organisation. WannaCry was highly contagious due to its wormish nature and ransomware attack required the payment of Bitcoin to unlock files. Cancellation of thousands of medical appointments and procedures in the National Health Service of the UK was one of the most severe effects. The effects of cyberattacks on transportation systems, manufacturing companies, and government agencies were significant around the world, demonstrating that cyberattacks may result in direct real-life effects. The failure of WannaCry was greatly attributed to the lack of an unpatched system, lack of proper network segmentation, and inadequate backup and recovery measures. The incident put into focus the paramount role played by prompt patch management and especially the systems that are connected to the internet.

The SolarWinds supply chain incident of 2020 has shown that nation-state actors are persistent and sophisticated. By infecting the SolarWinds Orion IT management software with malicious code, attackers broke into several government agencies and other privately owned organisations. This was a hacked software that was then sold to approximately 18,000 customers via normal update channels. The operational discipline demonstrated by the attackers was very high since the attackers chose particular targets to exploit them more, and in some cases had access to compromised systems over a period of months. They had tricks such as writing their own malware,

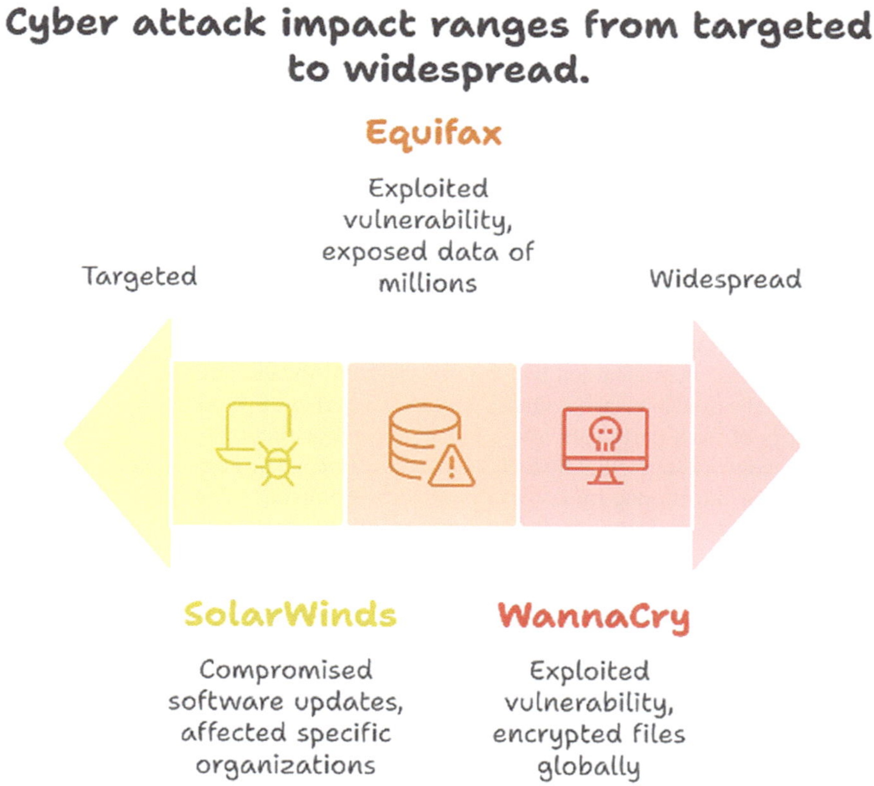

Fig. 8.4 Major cyber attacks

stealing legitimate credentials, and closely simulating normal network behaviour in order to evade detection.

In 2017, Equifax experienced one of the biggest personal information breaches in the history that disclosed the personal information of approximately 147 million Americans. The vulnerability that the attackers exploited in the Apache Struts framework had not been fixed, which shows the dangers of not addressing vulnerabilities quickly. The personal data, including Social Security numbers, date of birth, address, and driver's license information, was disclosed. The magnitude of the breach underscored the difficulty of keeping large datasets secure, exposing millions of individuals to long-term identity theft risks.

8.5.2 Lessons Learned from Major Incidents

The study of key cybersecurity events reveals general trends and systemic challenges that transcend the failures of particular organisations. Such tendencies can be used as valuable information to reinforce security measures and create more robust systems.

One of the common themes is that of the exploitation of vulnerabilities that have been known but have not been patched, and therefore, there has to be a powerful vulnerability management. Organisations should develop mechanisms for identifying, prioritising, testing, and implementing patches that meet security needs and ensure operational stability.

Another important lesson is the increasing significance of supply chain security. SolarWinds case demonstrated how trusted software updates and relationships with vendors can be used in massive attacks. Organisations have to be cautious in assessing the security of third-party software, services and infrastructure.

Lastly, network segmentation and good architecture would be extremely important to restrict the harm done by successful attacks. Companies that had well-isolated networks and had well-implemented access controls tended to better contain breaches than those with a flat interconnected network.

Incident response capabilities have a significant effect on the overall effect of security incidents. The reduction in the time of detection and recovery and overall impact is reduced, and organisations tend to have well-developed incident response plans, experienced personnel and regular exercises.

8.5.3 Success Stories: Zero Trust Adoption and Incident Response Excellence

The investigation of successful security deployments presents no less valuable information on effective practices and approaches, despite cybersecurity failures often making headlines. These success stories demonstrate that comprehensive cybersecurity strategies are effective in mitigating risks and minimising the impact of security incidents [16].

The implementation of zero-trust architectures has provided immense benefits to security in various organisations. One such case is the BeyondCorp initiative by Google, which abandoned the conventional use of VPN-based remote access and instead mandated validation of the user and device before each access attempt is made. This serves to decrease the risk of having open network connections all the time while enabling secure remote work.

Banks in the financial industry have implemented elaborate cybersecurity policies that align with corporate goals and regulatory standards. Most large banks now have sophisticated threat intelligence systems, real-time fraud detection systems and powerful incident response mechanisms in place to quickly detect and respond to threats.

Another way in which government agencies have modernised their cybersecurity activities is by patching vulnerabilities in old systems, implementing current security architectures, and developing advanced threat detection capabilities. These are some of the efforts that have enhanced their general cyber resilience.

8.6 Future Directions

8.6.1 Zero Trust Architectures

Zero-trust architectures have already demonstrated real benefits in the security of organisations. A notable example is the BeyondCorp program at Google, which no longer uses VPN-based access but instead requires device and user authentication with each request. This minimised the dangers of having continuous network links, and yet it maintained secure remote access.

The transition to the concept of zero trust indicates a larger trend of the abandonment of the old approach to security based on perimeter, which is no longer adequate in the modern context of cloud computing, mobile devices, and remote working. In this model, no user, device or location is trusted, each access request should continuously be checked based on factors such as identity, device health, and behaviour.

But it is not easy to practice zero trust. The challenges experienced by organisations include complex IT integration, ensuring utility and increased security, and the need to synchronise policies in various platforms.

8.6.2 Blockchain and Decentralised Security

Decentralisation of verification, immutable records, and cryptographic proofs are some of the new avenues in achieving trust in cybersecurity through blockchain. Although it does not cure everything, it can potentially be used in fields such as identity management, tracking the supply chain, and sharing secure data.

As an example, decentralised identity systems are based on blockchains that enable users to have greater control over their credentials and eliminate the risks of having a centralised identity database. Blockchain can create immutable records of component provenance, manufacturing processes, and delivery paths in a supply chain, making it simpler to trace down counterfeit components or un-authorised modifications.

8.6.3 Quantum Computing Risks and Opportunities

Quantum computing also poses serious threats and holds promise as far as cybersecurity is concerned. With the realisation of cryptographically capable quantum machines, the current public-key cryptography would cease to be secure. This would put a mass movement towards quantum resistant substitutes.

The problem is that the schedule of the practical quantum systems is unclear. However, since cryptographic updates require years to be implemented, the planning has to be started as early as possible. Current efforts in post-quantum cryptography are solving this with the creation of the algorithms that are resistant to the classical and quantum-attacks, but it is complicated to balance the performance, compatibility, and security.

8.6.4 Legal, Ethical, and Geopolitical Dimensions

Besides technical challenges, the law, ethics and geopolitics now have a greater impact on cybersecurity. Privacy laws such as GDPR and CCPA pose new demands in terms of data processing, notification of breaches and consent of users that directly affect risk management and security architecture.

Simultaneously, regulations on incident reporting, supply chain integrity and critical infrastructure defence are motivated by issues of national security. Cyber incidents are typically positioned in terms of national interest that influences the manner in which threats are apportioned, the manner in which information is disseminated, as well as the way in which international collaboration is sought.

8.7 Conclusion

8.7.1 Summary of Key Insights

This paper emphasises the fact that cybersecurity has become a critical, dynamic and rapidly evolving area. It is no longer about reactive perimeter defence but a proactive lifecycle-based security that is integrated into organisational operations.

The main principles of the CIA triad, risk management, and defence in depth are still central, but they can now be found within more frameworks, such as NIST and ISO 27001, which encourage more significant and uniform behaviour within sectors.

The threat environment is also evolving, and technological trends, such as cloud computing, IoT, and AI, as well as increased geopolitical and economic pressure, are impacting it. Such dynamics form new attack surfaces and require more innovative defence strategies.

In this context, penetration testing and ethical hacking have become a necessary part of security evaluation. The legitimacy of the researchers to enhance security output has also been established through the professionalisation of the field via legal frameworks, standards, and bug bounty programs.

8.7.2 Open Research Challenges

The research problems in cybersecurity are highly multidisciplinary, as they span multiple fields, including computer science, mathematics, psychology, economics, and policy. The advancements in automation and AI-based defences, as well as in architectural design, are needed to scale security solutions to ever-larger and more distributed systems.

The problem of human factors is still an uphill task, and it needs the wisdom of behavioural science and organisational studies. Adversarial AI presents two concerns; protecting against AI threats and protecting AI-based security systems. To be ready to the quantum era requires additional advancements in post-quantum cryptography and transition planning.

Homomorphic encryption, secure multi-party computation, and differential privacy are privacy-saving technologies that are becoming more and more topical. In the meantime, the economic models like cyber insurance and liability models need to be modified to adjust the incentives and responsibilities more.

Ultimately, technical innovation will not be the sole determinant of cybersecurity in the future; rather, this process will be shaped by social, ethical, and political forces. It takes long-term interdisciplinary cooperation to overcome these challenges so that cybersecurity could become one of the pillars of contemporary society instead of a field specialising in a niche.

References

1. Lnenicka M, Kysela T, Horák O. Building security and resilience: a guide to implementing effective cybersecurity and data protection measures in smart cities. Smart Sustain Built Environ. 2025;
2. Khanna A. Introduction to cybersecurity. In: Securing an enterprise: maximizing digital experiences through enhanced security measures. Berkeley, CA: Apress; 2025. p. 3–18.
3. Ramezan CA. Understanding the chief information security officer: qualifications and responsibilities for cybersecurity leadership. Comput Secur. 2025;152:104363.
4. Zacharis A, Gavrila R, Patsakis C, Douligeris C. Optimising AI models for intelligence extraction in the life cycle of cybersecurity threat landscape generation. J Inf Secur Appl. 2025;90:104037.
5. Nonum EO, Avwokuruaye O, Umar AM. Social engineering: understanding human factors in cyber security. Int J Converg Inform Sci Res. 2025;
6. Sardar TH, Pandey BK, editors. Machine learning hybridization and optimization for intelligent applications. 1st ed. CRC Press; 2024. https://doi.org/10.1201/9781003465775

7. Sardar TH, Pandey BK, editors. Big data computing: advances in technologies, methodologies, and applications. CRC Press; 2024.
8. Al Mahedi Hassan M, Fahim MF, Usman MM, Sardar TH, Mondal MA, Khan MA. Optimisation strategies for next-generation AI, ML, and IoT applications. In: Applications of computational learning and IoT in smart road transportation system. Cham: Springer Nature Switzerland; 2025. p. 163–84.
9. Usman MM, Fahim MF, Al Mahedi Hassan M, Sardar TH, Adhoni ZA, Gourisaria MK. Harnessing IoT and machine learning for sustainable, smart urban environments. In: Applications of computational learning and IoT in smart road transportation system. Cham: Springer Nature Switzerland; 2025. p. 199–214.
10. Bartko R, Kelemen R. Hackers in duty of cyberterrorism. J Infrastruct Policy Dev. 2025;9(2):10979.
11. Alfarizy MF, Zolkipli MF. The evolution of wireless penetration testing tools: a case study of aircrack-Ng and bettercap. Borneo Int J. 2025;8(2):40–7. ISSN 2636-9826
12. Ayala J, Ngo S, Garcia J. A deep dive into how open-source project maintainers review and resolve bug bounty reports. In: 2025 IEEE symposium on security and privacy (SP). IEEE; 2025. p. 522–38.
13. Khatun A, Sardar TH. Cryptography algorithms to prevent different security attacks. In: Next generation mechanisms for data encryption. CRC Press; 2025. p. 35–46.
14. Mohamed N. Artificial intelligence and machine learning in cybersecurity: a deep dive into state-of-the-art techniques and future paradigms. Knowl Inf Syst. 2025;30:1–87.
15. Singh T. Case studies: state-sponsored cyberattacks. In: Cybersecurity, psychology and people hacking. Cham: Springer Nature Switzerland; 2025. p. 151–65.
16. Pigola A, Meirelles FD. Zero trust in cybersecurity: managing critical challenges for effective implementation. J Syst Inf Technol. 2025;

Chapter 9
Internet of Things (IoT): Connecting the World

9.1 Introduction to IoT

The Internet of Things (IoT) represents a fundamental transformation in computing paradigms, where everyday physical objects become intelligent, interconnected entities capable of autonomous data collection, processing, and decision-making [1]. This revolutionary approach extends internet connectivity beyond traditional computing devices to encompass a vast ecosystem of embedded sensors, actuators, and smart objects that collectively form an intelligent infrastructure for the digital age [2].

9.1.1 *Background and Evolution of IoT*

The conceptual foundations of IoT emerged in the early 1980s with Carnegie Mellon University's internet-connected Coca-Cola machine, marking the first instance of an everyday appliance gaining network connectivity. However, the term "Internet of Things" was formally introduced by Kevin Ashton in 1999 during his work at Procter & Gamble, initially focusing on radio-frequency identification (RFID) applications in supply chain management.

The evolution of IoT has been driven by several converging technological trends. First of all, IoT has evolved to its current state due to the rapid development of fast and ever-connected communications enabled by broadband internet connectivity, wireless connectivity such as mobile networks and Wi-Fi in building areas, and 5G infrastructure. This evolution in communication has provided support for the IoT networks that depend on seamless device-to-device and device-to-cloud connectivity.

Sensor technology advancements have been equally critical, with technological miniaturisation and cost reduction enabling mass production of highly accurate, power-efficient sensors. Modern microelectromechanical systems (MEMS)

T. H. Sardar, *The Spectrum of Computer Science: Emerging Technologies and Trends*, Studies in Smart Technologies, https://doi.org/10.1007/978-3-032-19632-3_9

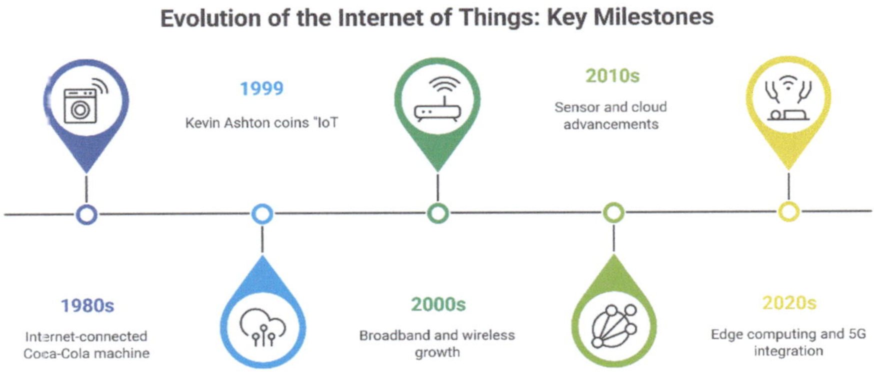

Fig. 9.1 The evolution of the internet of things (IoT) highlighting key milestones from the 1980s to the 2020s

technology facilitates detection of diverse physical phenomena including temperature, humidity, pressure, motion, light, sound, and chemical compositions at unprecedented scales and costs.

Cloud computing infrastructure provides the computational and storage backbone necessary to process massive volumes of IoT-generated data. Scalable cloud platforms accommodate dynamic IoT deployment requirements while big data analytics and machine learning capabilities transform raw sensor data into actionable insights for predictive maintenance, anomaly detection, and automated decision-making.

Miniaturisation and energy efficiency advances in semiconductor technology have enabled powerful yet energy-efficient system-on-chip (SoC) designs. Improved battery technologies and energy harvesting techniques using solar, thermal, and kinetic energy sources have extended device operational lifespans while reducing maintenance requirements.

The integration of edge and fog computing addresses latency, bandwidth, and privacy challenges inherent in centralised cloud processing by bringing computational capabilities closer to IoT devices. This distributed computing approach enables real-time processing and improved system responsiveness for applications requiring immediate response capabilities [3].

As shown in Fig. 9.1, the evolution of the Internet of Things (IoT) can be traced from early experiments such as the internet-connected Coca-Cola machine in the 1980s to the integration of edge computing and 5G technologies in the 2020s.

9.1.2 Significance of IoT in Today's World

IoT's significance extends beyond technological innovation to encompass fundamental transformations in operational efficiency, quality of life, business models, environmental sustainability, and societal structures.

The economic value of IoT is demonstrated by increased productivity and efficiency across industries. Predictive maintenance in manufacturing applications reduces unscheduled downtime by up to 30% while increasing equipment lifespans. Smart building systems usually achieve 15–25% energy savings by optimising energy consumption through automated HVAC control based on occupancy patterns. Real-time tracking features help supply chain management by lowering losses and enhancing inventory optimisation [4].

Improvements in consumer IoT applications that enhance quality of life include wearable technology that enables continuous health monitoring, home automation systems that offer convenience and energy efficiency, and smart city projects that improve urban living through better public services and optimised traffic management.

IoT-enabled servitisation, in which businesses shift from product-centric to service-centric offerings, has led to innovation in business models. Industrial equipment sold on a per-use basis, contracting based on results, and subscription-based connected device models are changing traditional business relationships and creating new sources of revenue based on data-driven services [5].

The positive aspects of environmental sustainability are accuracy in agriculture to maximise crop production, smart grid solutions to reduce energy use, and smarter transport networks that reduce emissions and traffic congestion by making the transport network efficient. With the increased use of IoT, several serious societal and ethical concerns should be carefully considered, including privacy protection, data ownership rights, algorithm bias, and the digital divide.

9.2 IoT Protocols and Architectures

Since IoT ecosystems are heterogeneous, complete architectural frameworks and standardised communication protocols are needed to ensure interoperability, scalability, and reliability over multiple deployment situations and requirements.

9.2.1 IoT Reference Architectures

Other well-known organisations have created reference architectures that address the different aspects of IoT system design and implementation [6].

Emphasising the role of security, privacy, and trustworthiness as key tenets of the Internet of Things, the National Institute of Standards and Technology (NIST) outlines a three-layer architecture that consists of device, communication, and application layers and incorporates security concerns on all tiers. The aim of the machine-to-machine (M2M) architecture developed by European Telecommunications Standards Institute (ETSI) is to standardise service-based communications protocols and interface using service-oriented meaning-making approaches. Standardised interfaces enable devices manufactured by different manufacturers to communicate with each other. Besides providing domain-specific solutions, which consider multiple application needs, the Internet of Things Architecture (IoT-A) project also came up with an extensive reference model whose scope covers functional, informational, communication, and security considerations. The four distinct layers into which the functionality is partitioned by the much-used layered stack approach are perception (sensors and actuators), network (communication protocols and infrastructure), middleware (data processing and management), and application (user interfaces and services). By clearly separating concerns and using standardised inter-layer interfaces, this modular approach makes design, implementation, and maintenance easier.

9.2.2 Overview of IoT Protocols

IoT protocols are a collection of various communication standards designed to meet the specific needs of IoT systems, including intermittent connectivity, power limitations, limited bandwidth, and heterogeneous device capabilities, while ensuring dependable data transfer and system interoperability [7].

Protocol selection depends on communication range, power consumption, data rate requirements, latency constraints, and cost considerations. Short-range protocols include Bluetooth Low Energy (BLE), Zigbee, and Wi-Fi suit local area networks and device-to-device communication. Medium-range protocols such as LoRaWAN and Sigfox address wide-area, low-power applications. Long-range protocols, including cellular technologies (4G, 5G, NB-IoT), provide extensive coverage and high data rates with higher power consumption and cost implications.

9.2.3 Key IoT Protocols

MQTT (Message Queuing Telemetry Transport) represents a lightweight, publish-subscribe messaging protocol designed for machine-to-machine communication in bandwidth-constrained environments. Operating on a broker-based architecture, MQTT enables devices to publish messages to specific topics and subscribe to relevant topics, creating flexible, scalable communication patterns. The protocol's three quality of service (QoS) levels provide different reliability guarantees: QoS 0 (at most once), QoS 1 (at least once), and QoS 2 (exactly once), enabling applications to balance reliability requirements with resource consumption.

CoAP (Constrained Application Protocol) specialises in web transfer for constrained devices and networks, standardised by the Internet Engineering Task Force (IETF). CoAP implements a request-response model similar to HTTP but optimised for low-power, lossy networks using UDP transport protocol. The protocol includes built-in support for multicast communication, service discovery, and resource observation, making it particularly suitable for sensor networks and building automation systems.

HTTP/HTTPS remains widely used in IoT applications with sufficient processing power and network bandwidth. HTTPS provides essential security through Transport Layer Security (TLS) encryption; however, computational and memory requirements can challenge resource-constrained devices, leading to the development of lightweight TLS variants and hardware-accelerated security solutions.

Zigbee and Z-Wave operate as IEEE 802.15.4-based wireless communication standards designed for low-power, low-data-rate applications. Through device-to-device communication relays, Zigbee's mesh networking topologies offer extended coverage and self-healing capabilities. To prevent Wi-Fi interference, Z-Wave is a proprietary wireless protocol for home automation that uses sub-gigahertz frequency bands.

By using a star-of-stars network topology and operating in unlicensed spectrum bands, LoRaWAN (Long Range Wide Area Network) offers low-power, wide-area connectivity with communication ranges of several kilometres in rural areas. While adaptive data rate (ADR) mechanisms optimise communication parameters based on link conditions, various device classes (Class A, B, and C) balance communication capabilities with power consumption.

5G and NB-IoT offer distinctive advances to the IoT with the ability to provide massive machine-type communication (mMTC) and support up to one million devices per square kilometre with ultra-low latency that is essential to address some critical phenomena. Narrowband IoT (NB-IoT) technology is a low-power wide-area network technology (LPWAN) that is provided in licensed cellular spectrum bands, which guarantees quality of service to applications with long battery life requirements and high coverage.

9.3 Security and Privacy in IoT

The nature of the IoT security and privacy challenges due to the specific features of the system, i.e. the restrictions of the resources, heterogeneous devices, and the necessity of ubiquitous data collection, and the subsequent need of multi-layered security strategies, the IoT security and privacy problems are an important challenge to consider in the context of ubiquitous IoT implementation [8].

9.3.1 IoT Security Challenges

Poor processing power, memory capacity, and access to energy means that IoT devices can do little with regard to employing robust security measures common to traditional computing systems. Security strength and consumption have to be balanced due to resource constraints and often, hardware-accelerated security solutions and lightweight security protocols have to be applied.

Due to the wide range of devices, protocols, platforms, and stakeholders involved in IoT implementations, diverse ecosystems present significant security challenges. When devices from various vendors interact within integrated systems, potential vulnerabilities may arise due to the differing security standards implemented by different manufacturers.

Inconsistencies in security implementation across platforms and vendors are caused by gaps in standardisation. IoT security guidelines have been developed by several standardisation bodies; however, the widespread adoption of unified security practices has been hindered by the voluntary nature of adoption and rapid technological advancements.

Numerous IoT devices positioned in exposed locations, where potential attackers could reach them, are susceptible to physical attacks. Physical access enables the direct extraction of sensitive data or cryptographic keys, facilitates side-channel analysis, and allows for hardware tampering, whereas remote deployment makes it challenging to promptly identify attacks.

IoT devices, whose operational periods are measured in years or decades during which security threats evolve significantly, are characterised by extended lifespans and update challenges. Economic models might not encourage continuous security support, and many devices are deployed in situations where regular updates are impractical or lack reliable update mechanisms.

Large volumes of sensitive and personal data are collected by IoT systems, raising concerns about privacy and data ownership. Although complex data flows involving multiple stakeholders make it challenging for users to understand data usage, granular and continuous data collection allows for detailed behavioural profiling that can be used beyond the original system objectives.

9.3.2 Common IoT Security Threats

Device hijacking is the unauthorised takeover of Internet of Things (IoT) devices, granting hackers access to private information, control over functionality, or the ability to utilise the devices as platforms for further attacks. Detection and remediation are particularly difficult due to the dispersed nature of IoT deployments and their limited monitoring capabilities.

Sensitive personal data, business intelligence, or system configuration data may be exposed to the public by IoT data breaches. Because data is continuously collected, breaches can use correlation and inference techniques to uncover historical patterns that reveal sensitive information.

To conduct distributed attacks, such as DDoS attacks, cryptocurrency mining, or spam distribution, botnets organise a sizable number of compromised IoT devices. As incidents like the Mirai botnet have demonstrated, IoT devices are particularly appealing targets for botnets due to their limited security capabilities and resource constraints.

Communications between IoT devices and their intended destinations can be intercepted and possibly manipulated by man-in-the-middle (MitM) attacks. Wireless protocols are susceptible to manipulation and interception, particularly when encryption is not used or is improperly configured.

Hardware tampering, cryptographic key extraction from device memory, and side-channel analysis are examples of physical attacks that use electromagnetic emissions, power consumption patterns, or timing variations to infer sensitive information.

Targeting Internet of Things devices, ransomware encrypts device data or interferes with functionality and demands payment to restore it. The potential operational and financial impact of critical infrastructure and industrial IoT systems makes them highly appealing targets, and their interconnectedness facilitates the rapid spread of attacks.

9.3.3 Security Solutions and Best Practices

Through cryptographic verification of boot loaders, operating systems, and applications, device security includes secure boot procedures that guarantee devices only launch trusted, authenticated software. Essential protection is provided by encryption of data in transit and at rest using simple algorithms designed for devices with limited resources. Hardware Security Modules (HSMs) offer secure execution environments for sensitive operations as well as tamper-resistant storage for cryptographic keys.

Access control and authentication. Implement multi-factor authentication (MFA), which requires multiple separate authentication factors before granting access to the system. Users and devices can only access resources that are suitable for their designated roles thanks to role-based access control, or RBAC. All system access must be continuously authenticated and authorised using thorough identity verification and network segmentation, according to zero-trust security models.

Network segmentation, which separates IoT devices from other resources and restricts their ability to move laterally in the event of a device compromise, is a key component of network security. Intrusion Detection and Prevention Systems (IDS/IPS) monitor device behaviour and network traffic for signs of malicious activity. Before allowing devices to access the network, Network Access Control (NAC) systems authenticate and authorise them.

End-to-end encryption is used in data protection to safeguard data at every stage of its lifecycle, including collection, transmission, storage, and processing. Data masking and anonymisation methods protect privacy by removing personally identifiable information whilst preserving data utility. By using the differential privacy techniques, statistical analysis can be done, allowing mathematical privacy.

Lifecycle management incorporates strong firmware update systems that help in the implementation of security patches within the entire life cycle of the device. Continuation of monitoring of the devices discovers potential malfunctions, abnormal performance, and breaches. Vulnerabilities in all components of the system are ensured to have been identified, analysed and fixed in a timely manner by undertaking comprehensive patch management practices.

Examples of human factors include programs of raising stakeholder awareness of threats and best practices in IoT security. Security by design principles are applied in the lifecycle of system development through incorporating security considerations into them. Constant security training makes sure that the processes are flexible enough to keep up with the changing threat conditions and technology.

9.4 IoT Applications and Use Cases

The flexibility and disruptive nature of IoT is demonstrated through a range of applications that solve particular issues and create new opportunities in different industries and spheres [9–11].

9.4.1 Smart Homes (Energy, Security, Convenience)

The fast-evolving Internet of Things applications allow smart home systems to combine various devices and services, which provide intelligent and responsive living environments that automate, optimise, and improve user experiences in residential environments.

Energy management systems are used to maximise the use of energy by the household through automated control of appliances, smart lighting controls and smart thermostats. To make smart heating, cooling and electrical usage decisions, advanced systems combine occupancy data, weather predictions and utility pricing information. Smart meter real-time consumption information assists homeowners to understand their consumption patterns and identify areas to achieve efficiency.

Security systems use motion sensors, security cameras, smart door locks, and video doorbells to offer multi-layered security and monitoring. Smartphone integration allows real-time notification and control of the system regardless of location. The artificial intelligence lowers the number of false alarms and keeps the protection effective by separating the regular operation and the possible threats.

Examples of convenience applications are voice-activated assistants that can be used as a central control interface (natural language control of appliances, entertainment and lighting). Smart appliances, with their remote monitoring and control capabilities and autonomous operation depending on user preferences and utilisation patterns, enable the user to tailor their living areas by automatically regulating the lighting, temperature and entertainment.

9.4.2 Wearable Technology (Health, Fitness, Safety)

By constantly and invisibly tracking physiological indicators and environmental factors, wearable IoT devices redefine personal health tracking, fitness and safety-related uses. They produce huge volumes of personal health information that can be used in proactive health management and emergency response.

Health monitoring involves the use of sensors which monitor vital signs such as the heart rate, blood pressure, oxygen levels of the blood and sleep patterns. High-tech wearable devices are capable of identifying irregular heartbeats, electrocardiogram (ECG) indicators and provide alerts to customers of a possible heart attack before it occurs. Continuous glucose monitoring can also help patients with diabetes to manage their condition and reduce the risk of developing complications by monitoring their blood sugar levels during the day.

Fitness tracking includes step counting, distance tracking, calorie estimation, and recognition of exercise. Up to date fitness trackers are also paired with nutrition trackers to manage health and wellness, can automatically recognise different kinds of physical activity, and provide personalised coaching based on performance and individual objectives.

The safety applications include environmental hazard monitoring, emergency location services, and senior citizen fall detection in the case of elderly people. Unless users react to predetermined timeframes, the fall detection features will automatically call to emergency services. Whereas the environmental sensors can detect the conditions that influence health, such as the UV radiation and the quality of the air, GPS tracking enables location-based safety services and emergency response aid.

9.4.3 Industrial IoT (Predictive Maintenance, Supply Chain, Automation, Digital Twins)

Industrial IoT (IIoT) deployments are advanced and cost-effective IoT applications that use sensors, analytics, and automation solutions to streamline industrial operations, lower the operational expenses, and enhance the quality of products in manufacturing, energy, and industrial industries.

In predictive maintenance systems, vibration sensors, temperature sensors, acoustic sensors and other instrumentation are used to continually monitor the state of equipment and predict the possibility of failures before they happen. Machine learning algorithms can be used to detect anomalies in sensor data patterns, which signal the development of a problem, and schedule the maintenance during a planned downtime instead of reacting to unexpected failures. Such systems extend the service life of equipment and minimise the unplanned downtimes by 30–50%.

Supply chain applications can be used with RFID tags, GPS tracking, and environmental sensors to provide real-time monitoring of supply chains, provide a clear end-to-end visibility of manufacturing processes, inventory quantities, and logistic operations. Sensitive products are protected by temperature and humidity monitoring, and demand forecasting and automated inventory management are made possible by ERP system integration.

Automation systems create highly automated manufacturing environments by combining IoT sensors with robotics, process control systems, and programmable logic controllers (PLCs). Human–machine interfaces provide comprehensive system visibility and control capabilities, while real-time sensor data enables dynamic adjustments to production parameters, ensuring optimal quality and efficiency.

Implementations of digital twins create virtual representations of real-world resources, procedures, or systems that are updated in real-time using data from IoT sensors. These digital models provide valuable insights for training, process optimisation, design validation, and scenario analysis, while enabling simulation, analysis, and optimisation without interfering with real-world operations.

9.4.4 *Smart Cities (Traffic, Environment, Safety, Waste Management)*

Smart city initiatives utilise IoT technologies to address urban issues such as public safety, traffic congestion, pollution, and municipal service delivery. These initiatives use comprehensive systems that integrate multiple IoT applications to create more sustainable, efficient, and livable urban environments.

To monitor traffic flow, identify congestion, and optimise signal timing, traffic management systems utilise sensors, cameras, and connected car technologies. Adaptive traffic control systems minimize the time spent at the road and fuel usage by changing the timing of signals based on the prevailing conditions. Parking management systems guide vehicles to vacant parking spots thereby alleviating traffic congestion and maximizing parking revenue by making use of it more conveniently.

Environmental monitoring networks place sensors all over the cities to monitor environmental parameters, noise levels, water quality, and air quality. Real-time environmental data can help the officials to identify the causes of pollution, assess the effectiveness of environmental policies, and make public health warnings where necessary. Smart street lighting is energy-saving and does not compromise on the safety of the people since it can adjust the light depending on use patterns and weather conditions.

Public safety Applications in public safety can encompass crime prediction analytics, emergency response coordination, and intelligent video surveillance. IoT-powered surveillance cameras allow law enforcement to respond faster because they are able to detect abnormal behavior or security threats automatically. Integrated data sources enable predictive policing methods, and gunshot detection systems facilitate locating and identifying gunfire and reduce the response time to an emergency.

Waste management systems also optimise collection routes and schedules, as opposed to using preset schedules, on the basis of actual waste levels. Smart waste bins with fill-level sensors enable dynamic collection vehicle routing, which saves labour and fuel costs without compromising the quality of the services. Recycling monitoring indices are used to track recycling levels where waste minimisation efforts are to be improved.

9.4.5 Agriculture and Farming (Precision, Livestock, Traceability)

Precision agriculture methods optimise resource use while enhancing crop yields and quality, and agricultural IoT applications help meet the growing demand for food production worldwide, encouraging environmentally responsible farming methods and sustainable farming practices.

Precision agriculture systems provide comprehensive field condition, crop health, and growth pattern data by utilising soil sensors, weather stations, satellite imagery, and drone-based monitoring. By enabling precise irrigation, fertilisation, and pesticide control based on particular field conditions and crop requirements, variable rate application systems can lower input costs by 10–20% while preserving or increasing yields.

Livestock monitoring systems improve herd management and animal welfare by tracking the location, health, and behavioural patterns of animals. While automated feeding offers customised nutrition, wearable sensors track vital signs, activity levels, and reproductive status to enable early detection of health problems and optimal breeding management.

Traceability systems provide transparency on the production process, handling technique, and quality characteristics of agricultural products as they are tracked from farm to consumer. Blockchain-based traceability enables premium pricing for sustainably produced goods, while generating immutable product history records that facilitate the quick identification and containment of food safety issues.

9.4.6 Healthcare IoT (IoMT) (Remote Monitoring, Smart Hospitals, Pandemic Response)

The Internet of Medical Things (IoMT) encompasses IoT applications specifically designed for patient monitoring, medical device management, and healthcare delivery. Through effective and efficient care delivery models, IoMT has the potential to significantly improve patient outcomes while lowering healthcare costs.

Outside of conventional clinical settings, remote patient monitoring systems enable medical professionals to track patients' vital signs, medication adherence, and overall health. While telehealth platforms integrate with IoMT devices to provide comprehensive remote care capabilities, chronic disease management programs utilise IoMT devices to monitor conditions such as diabetes, hypertension, and heart disease, enabling early intervention and preventing complications that may require hospitalisation.

IoT technologies are integrated into smart hospital systems to enhance patient safety, operational effectiveness, and care quality. Asset tracking systems monitor the location and condition of medical equipment, reducing device searches and ensuring the availability of vital equipment. While environmental monitoring guarantees proper conditions in crucial areas, patient flow management tracks movement through hospital facilities, optimising resource allocation and cutting wait times.

Applications for pandemic response showed how important IoT is to managing public health during COVID-19. Contact tracing apps utilised Bluetooth and smartphone sensors to detect potential exposure events and promptly notify at-risk individuals. While ventilator management systems enabled remote monitoring and control of vital care equipment, temperature screening systems provided non-invasive health screening.

9.4.7 Automotive IoT (V2X, Safety, Fleet Management)

By combining sensors, connectivity, and artificial intelligence, automotive IoT applications revolutionise mobility through improved safety systems, increased efficiency, and the development of autonomous vehicle capabilities.

Vehicles can communicate with other vehicles (V2V), networks (V2N), pedestrians (V2P), and infrastructure (V2I), thanks to vehicle-to-everything (V2X) communication. V2V communication allows information sharing about location, speed, direction, and braking status enabling collision avoidance and cooperative driving behaviours. V2I communication provides real-time traffic condition, road hazard, and optimal routing information demonstrating 25–40% traffic delay reductions and significant accident reduction potential.

Advanced Driver Assistance Systems (ADAS) offer safety features such as adaptive cruise control, lane keeping assistance, automatic emergency braking, and blind spot monitoring, utilising cameras, radar, lidar, and ultrasonic sensors. While V2X communication integration expands capabilities beyond onboard sensor ranges, these systems continuously monitor the environment around the vehicle and can take action to prevent or lessen collisions.

Fleet management systems provide comprehensive monitoring and optimisation for commercial vehicle operations through GPS tracking, engine diagnostics, driver behaviour monitoring, and fuel consumption analysis, enabling route optimisation, operating cost reduction, and safety performance improvement. Predictive maintenance systems reduce costs and minimise vehicle downtime through optimal maintenance scheduling.

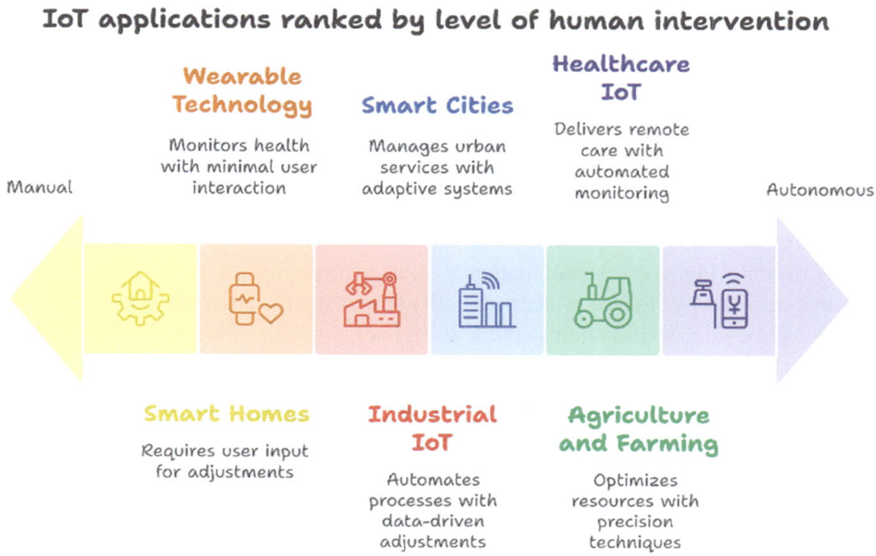

Fig. 9.2 IoT applications ranked by level of human intervention

As illustrated in Fig. 9.2, IoT applications span a spectrum of human involvement, beginning with manual adjustments in smart homes and progressing toward autonomous systems in healthcare IoT. This continuum highlights how increasing automation is reshaping domains such as wearables, smart cities, industrial IoT, and agriculture.

9.4.8 Retail and Energy IoT (Smart Shelves, Smart Grids)

The adoption of IoT in the retail and energy sectors enhances customer experiences, improves operational efficiency, and unlocks business models that leverage real-time data and analytics. RFID-tagged inventory management, weight-sensor-equipped smart shelves, and computer vision systems that automatically monitor product availability and consumer purchasing patterns are examples of smart retail applications. Automated reordering and dynamic pricing are made possible by real-time inventory monitoring. Augmented reality and smart mirrors enhance customer shopping experiences by providing virtual try-on capabilities, product information, and personalised recommendations.

IoT sensors are used by customer analytics systems to monitor dwell times, product interactions, and customer movement patterns. While facial recognition and emotion detection offer insights into customer preferences and satisfaction, heat mapping pinpoints high-traffic areas and the best places for products, allowing retailers to customise marketing messages and optimise store layouts.

IoT technologies are integrated into electrical power systems through smart grid systems, which enhance sustainability, reliability, and efficiency. Demand response programs and time-of-use pricing are made possible by smart meters' real-time consumption data, which lowers grid stress and peak demand. By identifying and isolating faults, distribution automation reduces outages and enhances service reliability.

IoT monitoring and control enhance the performance of solar panels, wind turbines, and energy storage systems, thereby benefiting the integration of renewable energy. While microgrid systems enable local energy optimisation and grid independence during outages, grid-scale battery management systems monitor performance, optimise charging cycles, and extend operational lifespans.

9.5 Future Trends and Research Directions

Understanding new trends and research directions is crucial to predict future developments and prepare for next-generation IoT capabilities, as the evolution of IoT technology continues to accelerate due to advancements in artificial intelligence, edge computing, quantum technologies, and sustainable computing approaches [12–14].

9.5.1 *AI/ML-Enabled IoT (Edge Intelligence)*

A fundamental change from centralised cloud-based analytics to distributed intelligence is represented by the direct integration of AI and machine learning into IoT devices and edge computing infrastructure. By processing sensitive data locally rather than sending it to distant servers, edge AI enhances privacy protection, reduces latency and bandwidth requirements, and enables real-time decision-making.

Increasingly powerful embedded AI processors with low power consumption are being developed for battery-operated devices, making them suitable for Internet of Things applications. Complex machine learning models on devices with limited resources are made possible by these processors' optimised neural network inference architectures. Federated learning techniques protect privacy while increasing model accuracy by facilitating cooperative model training across dispersed devices without centralising sensitive data.

Applications of edge intelligence include autonomous car systems that need to make instantaneous decisions based on sensor data without relying on network connectivity, intelligent traffic management that responds instantly to changing conditions, and real-time anomaly detection in industrial systems.

9.5.2 Blockchain for IoT Security and Data Integrity

With their decentralised trust mechanisms, immutable data records, and smart contract capabilities, blockchain technologies present viable answers to IoT security and trust issues. Integrating blockchain technology with Internet of Things systems can enhance security, unlock new business opportunities, and provide transparent audit trails for data transactions and device behaviour.

Device identity management is a promising blockchain application that offers decentralised authentication and identity verification without the need for centralised authorities. Trustworthy device authentication across various ecosystems is made possible by blockchain-based device registries, which keep tamper-proof records of credentials, software versions, and security status.

Blockchain technology is used in supply chain transparency to create immutable records of the manufacturing, handling, and transportation processes. This enables quick identification and response to quality or safety issues, while providing consumers with information about authentic, sustainable, and high-quality products.

9.5.3 Quantum-Resilient IoT Systems

The development of quantum-resilient security mechanisms for Internet of Things systems is necessary due to the potential for quantum computers to crack existing cryptographic systems. The aim of post-quantum cryptography is to open encryption algorithms that resist both classical and quantum computing attacks. Quantum-resistant cryptographic algorithms are very challenging to implement in the resource-constrained IoT devices because of their increased computational needs and the larger key size in comparison to current methods. With the emergence of quantum computing, it is important to research lightweight post-quantum cryptographic algorithms that are specifically designed to support Internet of Things applications in order to maintain security in the long term.

9.5.4 Digital Twin–IoT Integration

With the convergence of IoT systems and digital twin technology, it forms an all-encompassing virtual representations of real-world objects, processes, and places that is constantly updated in real-time with sensor data. This integration enables advanced simulation, analysis and optimisation capabilities reducing operational risks and improving system performance.

Advanced digital twin implementations that combine artificial intelligence and machine learning make predictive modelling, anomaly detection, and automated decision-making through analysis of virtual models possible. These systems allow experimentation and optimisation without any risk by simulating suggested changes or operational conditions without any effect on physical systems.

City-scale digital twins combine the data of many IoT systems, including public safety, energy management, environmental sensing, and traffic monitoring, into comprehensive urban models that can be used to develop a policy, coordinate responses during emergencies, and plan the city.

9.5.5 *Sustainability and Green IoT*

The development of green IoT technology is driven by environmental sustainability issues, which seek to reduce environmental impact by managing the lifecycle of devices in a circular economy, using sustainable materials, and using less energy. IoT devices can utilise ambient energy sources, such as solar, thermal, kinetic, and radio frequency energy, thanks to energy harvesting technologies. By extending the operational lifespans of battery-powered devices and eliminating the need for batteries in certain applications, these methods can lower maintenance costs and electronic waste.

The development of IoT device components that safely decompose at the end of their useful lives is the main goal of research into biodegradable and sustainable materials, which also aims to reduce the accumulation of electronic waste. Longer lifespans and lower resource consumption are made possible by modular device designs, which enable component reuse and upgrades.

9.6 Case Studies

Practical challenges, advantages, and lessons learnt from extensive deployments across various application domains and geographic contexts are all illustrated by real-world IoT implementations, which show both opportunities and barriers in IoT adoption [15–18].

9.6.1 *Smart Cities Implementation*

Barcelona's comprehensive smart city initiative integrates multiple systems improving city operations and citizen services. The city deployed over 20,000 smart water meters providing real-time consumption data and leak detection, reducing water waste by 15% while improving service reliability. The smart parking system utilises ground-embedded sensors monitoring space availability in real-time, reducing traffic congestion from parking searches by approximately 30% while increasing parking revenue. Environmental monitoring includes 500 air quality sensors distributed throughout the city providing real-time pollution data enabling targeted interventions and policy decisions.

Singapore's Smart Nation initiative encompasses comprehensive IoT deployments across transportation, healthcare, urban planning, and government services. The traffic management system integrates thousands of sensors, cameras, and GPS devices providing real-time optimisation and incident detection while dynamic road pricing adjusts tolls based on conditions, reducing congestion and generating infrastructure revenue. The national sensor platform integrates environmental monitoring throughout the island tracking air quality, water quality, noise levels, and weather conditions supporting public health protection and environmental policy development.

9.6.2 Industrial IoT Deployment

General Electric's Predix platform demonstrates industrial IoT potential for data-driven operational optimisation across aviation, power generation, and healthcare industries. Aviation applications analyze thousands of aircraft engine sensors predicting maintenance requirements and optimizing flight operations, enabling airlines to reduce unscheduled maintenance by 35% while improving fuel efficiency. Power generation applications optimise wind turbine, gas turbine, and generation asset performance, reducing downtime by 25% while increasing power output by 3–5%.

Siemens MindSphere platform provides comprehensive industrial IoT capabilities for manufacturing, building technologies, and mobility applications, integrating data from diverse industrial systems including PLCs, sensors, and enterprise systems for unified analytics and optimisation. Manufacturing applications demonstrate 10–20% productivity improvements through optimised production scheduling, quality management, and predictive maintenance while energy management applications achieve 15–25% energy consumption reductions.

As shown in Fig. 9.3, the future of IoT will be shaped by advancements in AI, digital twins, and edge computing. These trends emphasise interoperability, security, and sustainability as the foundation for next-generation IoT ecosystems.

9.7 Conclusion

9.7.1 Summary of Key Insights

The Internet of Things has evolved from a conceptual framework into a transformative technology that is reshaping industries, cities, and daily life. This in-depth analysis reveals several key lessons about the current state of the IoT and its potential.

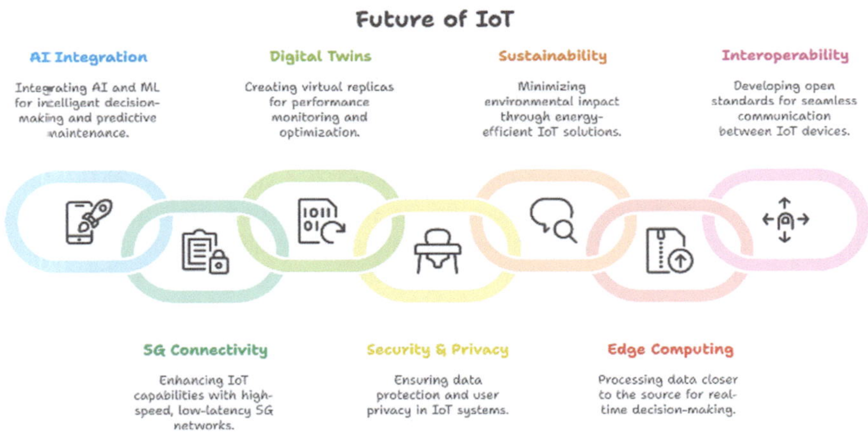

Fig. 9.3 The future trends and research directions in internet of things

These technological building blocks have come of age, owing to the development of sensor technology, communication protocols, and computational power, which can be used to develop advanced applications in various fields. The combination of cloud computing, edge processing, and artificial intelligence enables the development of IoT systems that can perform autonomous tasks and make intelligent decisions, something previously impossible.

The variety of IoT applications proves the versatility of technology and a wide range of use, including smart homes that make homes more comfortable and efficient or industrial systems that streamline intricate manufacturing processes. The economic effect is ever-increasing with industry analysts estimating further growth in all the key areas of application.

The issue of security and privacy is still a major concern that should be addressed and innovation should occur constantly. The peculiarities of the IoT systems such as resource limitations, the variety of deployment locations, and extended operating periods bring complicated security issues that cannot be resolved in full by conventional methods of cybersecurity.

9.7.2 Challenges Ahead

Although there have been considerable advancements, there are still a number of issues to resolve to ensure the full potential of the IoT systems is realised, and these are across technical, economic, regulatory, and social aspects and something that requires concerted action by stakeholders.

The technical issues involve creating more efficient and secure communication protocols, better battery technologies and energy harvesting, and better artificial intelligence algorithms that have been optimised to run on resource-constrained environments. The combination of new technologies, such as 5G networks, edge computing, and blockchain systems, introduces new complexity which needs to be managed effectively.

Economic challenges involve developing sustainable business models that support long-term device and service maintenance, particularly for applications with extended operational lifespans. The total cost of ownership must account for deployment, operation, maintenance, and end-of-life management costs that span decades.

Regulatory challenges encompass privacy protection, cybersecurity requirements, spectrum management, and international coordination issues. The global nature of IoT applications requires international cooperation and harmonisation of regulatory approaches, enabling seamless cross-jurisdictional operations.

9.7.3 Open Research Questions

The continued evolution of IoT technology raises numerous research questions that shape future development and deployment, spanning fundamental technical questions, application-specific optimisations, and broader systemic issues.

Fundamental technical research encompasses the development of ultra-low-power computing architectures that support sophisticated AI algorithms on battery-powered devices, quantum-resistant cryptographic protocols optimised for IoT constraints, and novel communication protocols that adapt dynamically to changing network conditions.

Application-specific research focuses on optimising IoT systems for particular domains, including healthcare IoT research that addresses sensor accuracy, data privacy protection, and clinical workflow integration. Meanwhile, industrial IoT research investigates advanced predictive analytics and human–machine interaction in automated environments.

The interdisciplinary nature of IoT research requires collaboration between computer scientists, engineers, social scientists, economists, and policymakers to address complex challenges comprehensively. Solutions developed in response will determine whether IoT technologies fulfil their transformative potential while addressing legitimate concerns regarding privacy, security, and social impact.

Future IoT success depends on the thoughtful integration of technology into social, economic, and environmental systems, enhancing human welfare while preserving essential values, including privacy, security, and equity. Continued IoT development success requires maintaining a balance between technological capability and human-centred design principles, ensuring technology serves humanity's broader goals and aspirations.

References

1. Sahana DS, Sardar TH, Sureddy S. Future intelligent vehicles: research roadmaps, open issues, and key challenges. In: Applications of computational learning and IoT in smart road transportation system. Cham: Springer Nature Switzerland; 2025. p. 1–16.
2. Mallik S, Qin H, Nandi S, Alam MY, Roy A, Sardar TH, editors. Applications of computational learning and IoT in smart road transportation system. Springer Nature; 2025.
3. Fahim MF, Usman MM, Al Mahedi Hassan M, Sardar TH, Bindu Madavi KP, Venkatachalam N. Revolutionizing road transportation: the role of artificial intelligence in smart and efficient systems. In: Applications of computational learning and IoT in smart road transportation system. Cham: Springer Nature Switzerland; 2025. p. 87–107.
4. Bose A, Sardar TH, Mridha SK. Recent advancements and future perspectives of dynamic fuzzy controllers for smart traffic signaling. In: Applications of computational learning and IoT in smart road transportation system. Cham: Springer Nature Switzerland; 2025. p. 109–28.
5. Bindu Madavi KP, Krishna Sowjanya K, Sardar TH, Reddy MS, Nimmalapudi SR. Smart traffic systems: revolutionizing road transport with AI and image processing. In: Applications of computational learning and IoT in smart road transportation system. Cham: Springer Nature Switzerland; 2025. p. 185–98.
6. Di Martino B, Rak M, Ficco M, Esposito A, Maisto SA, Nacchia S. Internet of things reference architectures, security and interoperability: a survey. Internet Things. 2018;1(1):99–112.
7. Mansour M, Gamal A, Ahmed AI, Said LA, Elbaz A, Herencsar N, Soltan A. Internet of things: a comprehensive overview on protocols, architectures, technologies, simulation tools, and future directions. Energies. 2023;16(8):3465.
8. Chanal PM, Kakkasageri MS. Security and privacy in IoT: a survey. Wirel Pers Commun. 2020;115(2):1667–93.
9. Usman MM, Fahim MF, Al Mahedi Hassan M, Sardar TH, Adhoni ZA, Gourisaria MK. Harnessing IoT and machine learning for sustainable, smart urban environments. In: Applications of computational learning and IoT in smart road transportation system. Cham: Springer Nature Switzerland; 2025. p. 199–214.
10. Al Mahedi Hassan M, Fahim MF, Usman MM, Sardar TH, Mondal MA, Khan MA. Optimisation strategies for next-generation AI, ML, and IoT applications. In: Applications of computational learning and IoT in smart road transportation system. Cham: Springer Nature Switzerland; 2025. p. 163–84.
11. Suresh A, Nandagopal M, Raj P, Neeba EA, Lin JW. Industrial IoT application architectures and use cases. Auerbach Publications; 2020.
12. Alnumay WS. The past and future trends in IoT research. Natl J Antennas Propag. 2024;6(1):13–22.
13. Shafique K, Khawaja BA, Sabir F, Qazi S, Mustaqim M. Internet of things (IoT) for next-generation smart systems: a review of current challenges, future trends and prospects for emerging 5G-IoT scenarios. IEEE Access. 2020;28(8):23022–40.
14. Ghosh A, Edwards DJ, Hosseini MR. Patterns and trends in internet of things (IoT) research: future applications in the construction industry. Eng Constr Arch Manag. 2021;28(2):457–81.
15. Muntjir M, Rahul M, Alhumyani HA. An analysis of internet of things (IoT): novel architectures, modern applications, security aspects and future scope with latest case studies. Int J Eng Res Technol. 2017;6(6):422–47.
16. Patti E, Acquaviva A. IoT platform for smart cities: requirements and implementation case studies. In: 2016 IEEE 2nd international forum on research and technologies for society and industry leveraging a better tomorrow (RTSI). IEEE; 2016. p. 1–6.
17. Alladi T, Chamola V, Sikdar B, Choo KK. Consumer IoT: security vulnerability case studies and solutions. IEEE Consum Electron Mag. 2020;9(2):17–25.
18. Schimek RS. IoT case studies: companies leading the connected economy. American International Group; 2016. p. 1–6.

Chapter 10
Cloud Computing and Virtualisation

10.1 Introduction

Cloud computing is one of the most radical concepts in modern IT, and it has entirely changed the way business operations deal with data, application deployment, and the use of computer resources. The game changer in the move toward distributed, scalable cloud computing over traditional on-premises computing has now made enterprise-grade computing capabilities more accessible. This has enabled hitherto unexplored degrees of versatility, effectiveness, and novelty in various fields and industries.

10.1.1 Background and Evolution of Cloud Computing

The conceptual foundations of cloud computing can be traced to the 1960s, when computer scientist John McCarthy forecasted that eventually computing would be organised as a civic utility, just like water and electricity. It is this concept of computing as a service that formed the basis of the present paradigm of cloud computing [1].

Cloud computing has been through various phases of development that have been sequentially based on development in technology and the changing business demands. The time-sharing systems of the mainframe era of the 1960s and 1970s were the early resource sharing examples where the computer resources could be accessed by hundreds of users simultaneously. The client–server architecture, which was developed in the 1980s and 1990s, was the first architecture to introduce concepts of distributed computing that would eventually influence the development of cloud architectures.

The result of the dot-com boom of the late 1990s and early 2000s was that companies such as Amazon ended up with internal infrastructure capabilities that far

T. H. Sardar, *The Spectrum of Computer Science: Emerging Technologies and Trends*, Studies in Smart Technologies, https://doi.org/10.1007/978-3-032-19632-3_10

exceeded their short-term needs, and that resulted in an unprecedented demand on scalable computing resources.

This surplus, supported by the development of the virtualisation technology and the high speed of its internet connection, formed the environment in which the commercial cloud services could have emerged.

Launched in 2002 as a simple web services offering, Amazon Web Services, which was expanded in 2006 with the introduction of Elastic Compute Cloud (EC2), is generally credited with bringing cloud computing out of the realm of an academic theory and making it a feasible commercial reality. The popularity of AWS showed that the industry needed on-demand, pay-as-you-use computing resources and motivated the creation of competing offerings by Microsoft, Google, and other large technology firms.

Figure 10.1 traces the conceptual foundations and evolution of cloud computing, beginning with John McCarthy's 1960s prediction of computing as a public utility, followed by early phases such as Mainframe (1960s-1970s), with multiple users accessing resources simultaneously, and Client–Server Architecture (1980s-1990s), which influenced future cloud designs. The dot-com boom (late 1990s-early 2000s) drove demand for scalable resources, supported by Amazon's internal infrastructure and key technological enablers like virtualisation and high-speed internet, culminating in the emergence of commercial cloud services with AWS's launch in 2006, introducing the Elastic Compute Cloud (EC2) and a pay-as-you-use model that catalysed the market demand for on-demand infrastructure

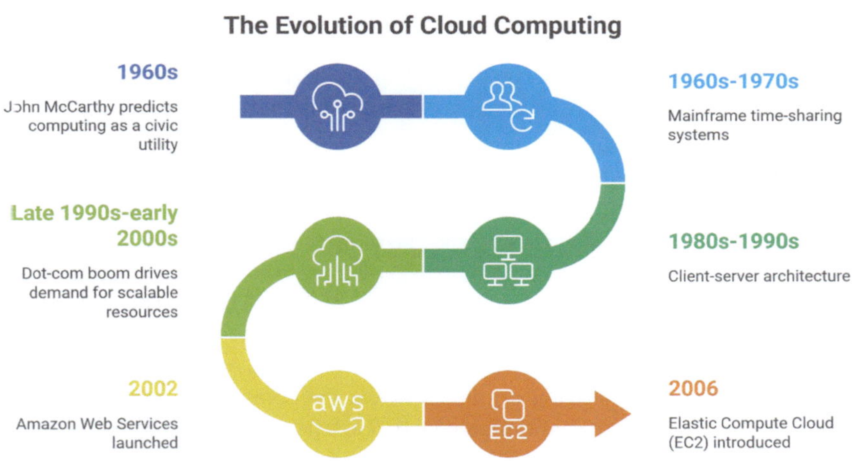

Fig. 10.1 Evolutionary timeline of cloud computing concepts

10.1.2 Importance of Virtualisation in Cloud Paradigm

Virtualisation technology serves as the foundational enabler of cloud computing, providing the abstraction layer that allows physical hardware resources to be dynamically allocated, shared, and managed across multiple users and applications. Without virtualisation, the elastic, on-demand characteristics that define cloud computing would be impossible to achieve at scale [2].

Beyond just resource sharing, virtualisation is crucial for cloud computing because it offers vital features like workload isolation, resource optimisation, and dynamic scalability. Virtualisation enables cloud providers to get maximum hardware utilisation by running many virtual instances on a single physical infrastructure, and ensuring security and performance isolation between tenants.

The key element of virtualisation systems is the hypervisor technology, which creates virtual machines, which even operate like different computers despite the fact that they are using the same hardware facilities. This abstraction makes possible the automation and orchestration of large-scale cloud workload operations, as cloud providers can provide standardised computing environments, which can be provisioned, configured and managed programmatically.

10.1.3 Objectives and Structure of the Chapter

This chapter gives a thorough review of cloud computing and virtualisation technologies, its implementation strategies, technical principles, and practical use in a number of areas. The primary objectives are to learn the fundamentals of cloud computing, examine virtualisation technologies and how they facilitate the provision of cloud services, evaluate security, privacy, and governance challenges, examine leading cloud service vendors and products, and explore new trends and directions.

Starting with the basic concepts through to technical applications, to practical applications and future outlook, the structure of the chapter flows in logical sequence. Using this approach, the readers will have the theoretical understanding and practical skills needed to achieve a successful adoption and implementation of cloud computing.

10.2 Fundamentals of Cloud Computing

Cloud computing is a paradigm shift, where computing resources are no longer spent on capital expenditure on physical infrastructure, but as a service that is consumed on-demand. To value the opportunities and limitations of cloud computing deployments, one needs to understand the fundamental aspects of service model, basic features of cloud computing, and cloud deployment strategies.

10.2.1 Characteristics of Cloud Computing

The National Institute of Standards and Technology (NIST) has developed five major features that distinguish cloud computing as compared to the traditional methods of computing. Combined, these characteristics constitute the unique value proposal that has given rise to the widespread adoption of cloud services in a variety of industries and applications.

On-demand self-service enables the customers to automatically add and supply computer resources without the direct interaction with the service providers. This capability can make organisations respond promptly and eliminate the old types of procurement and administrative overheads. Programmatic resource management and automation are provided via application programming interfaces (APIs); user-friendly interfaces are provided by self-service portals to provide resource provisioning, configuration management, and monitoring.

Cloud services are accessible by a broad range of client devices such as laptops, tablets, smartphones, and workstations, and are reachable by broad network access, meaning that they are accessible over standard networks using standard protocols. Conventional on-premises infrastructure would complicate or prevent the ability to adopt flexible working hours, international collaboration, and mobile workforces.

Resource pooling can also allow cloud providers to consolidate computing resources and provide them to many customers through multi-tenant models that optimise resource utilisation and maintain security and performance isolation. This pooling strategy, which reduces costs and provides access to enterprise-grade infrastructure, makes economies of scale possible and otherwise uneconomical to individual organisations.

Due to the speed of elasticity, the provisioning and release of computing resources is automatically scaled to meet demand patterns. This ability will enable organisations to manage the unpredictable workloads associated with variable demand and avoid unnecessary over- and under-provisioning of infrastructure when demand is low and high, respectively. Elasticity may be used on any number of time scales, including planned scaling (when the demand pattern is expected to operate) and automatic scaling (when the demand pattern actually varies in real time).

Measured services make resource optimisation and usage-based billing models possible, with their clear monitoring and reporting of resource usage. Metering systems enable a detailed understanding of the trends of resource utilisation as they track the utilisation of computing resources, storage capacity, and network bandwidth among other measurable parameters.

10.2.2 Cloud Service Models

Cloud computing services are usually categorised into three key service models that indicate different levels of abstraction and accountability on the part of the manager. To select the appropriate cloud services and develop effective cloud adoption strategies, one needs to understand such models [3].

The primitive computing resources provided in the Infrastructure as a Service (IaaS) model include virtual machines, storage systems, networks, and other infrastructure components on a pay-as-you-use basis. IaaS solutions save capital expenditure and management costs on physical infrastructures and provide the greatest flexibility and control of computing environments. Although cloud vendors deal with physical infrastructure, virtualisation layer and simple network connectivity, the Infrastructure as a Service (IaaS) consumers must manage the operation system, application deployment, security setting, and data management. Some of the major IaaS available comprise Amazon EC2, Microsoft Azure Virtual Machines, Google Compute Engine, and IBM Cloud Virtual Servers.

PaaS provides application development tools, libraries, and services optimised around application development, development and deployment environments, which abstract the underlying infrastructure complexity. PaaS platforms also enable developers to focus on business functionality and application logic, instead of handling and maintaining infrastructure tasks. Elements that are typically found in PaaS products are database management systems, middleware services, application servers, deployment automation tools, and integrated development environments. AWS Elastic Beanstalk, Microsoft Azure App Service, Google App Engine, and Heroku are well-known PaaS platforms.

Software as a Service (SaaS) delivers complete applications over the internet, eliminating the need for organisations to install, configure, and maintain software on local systems. SaaS applications are typically accessed through web browsers or thin-client applications, providing ubiquitous access to business functionality without requiring local software installations.

Leading SaaS applications include customer relationship management systems, such as Salesforce, productivity suites like Microsoft Office 365 and Google Workspace, and enterprise resource planning systems, like SAP SuccessFactors. Figure 10.2 compares cloud service models: IaaS (Infrastructure as a Service) provides infrastructure resources with customer responsibility for OS, apps, security, and data, and examples include Amazon EC2 and Azure VMs; PaaS (Platform as a Service) offers development and deployment environments with customer focus on application development and management, exemplified by Heroku and Google App Engine; SaaS (Software as a Service) delivers complete applications over the internet with no customer responsibility, relying on provider-hosted solutions like Salesforce and Office 365, as of August 2025.

10.2.3 Cloud Deployment Models

Cloud deployment models define how cloud infrastructure is provisioned, managed, and accessed by different types of organisations and user communities. The choice of deployment model depends on factors including security requirements, compliance

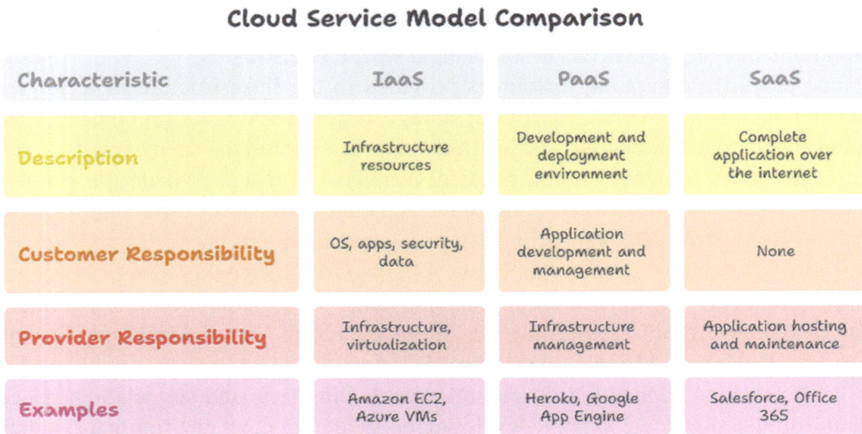

Fig. 10.2 Comparative overview of cloud service models

obligations, cost considerations, and organisational preferences for control versus convenience [4].

Public Cloud services are provided by third-party vendors who own and operate cloud infrastructure that serves multiple customers through shared resources. Public clouds offer the greatest economies of scale, the highest levels of elasticity, and the lowest barriers to entry, making them attractive for organisations seeking cost-effective access to enterprise-grade computing capabilities.

Private Cloud infrastructure is dedicated to a single organisation, providing cloud computing benefits while maintaining greater control over security, compliance, and resource allocation. Private clouds can be hosted on-premises, by third-party providers, or through hybrid arrangements that combine multiple hosting approaches.

Hybrid Cloud architectures combine public and private cloud resources to create integrated environments that leverage the benefits of multiple deployment models. Hybrid approaches enable organisations to optimise workload placement based on security requirements, performance needs, cost considerations, and regulatory constraints.

Community Cloud infrastructure serves specific communities of organisations with shared requirements, compliance obligations, or security concerns. These specialised environments enable cost sharing while maintaining tighter controls than public clouds and greater economies of scale than individual private clouds. Figure 10.3 compares cloud deployment models including Public Cloud (third-party ownership, shared resources, least control, cost-effective for economies of scale), Private Cloud (single organisation ownership, dedicated infrastructure, greater control, potentially expensive for security/compliance), Hybrid Cloud (combination ownership, integrated environments, variable control, optimised workload placement), and Community Cloud (specific community ownership, shared infrastructure, tighter control, cost sharing for shared requirements).

Cloud Deployment Models Comparison

Characteristic	Public Cloud	Private Cloud	Hybrid Cloud	Community Cloud
Ownership	Third-party vendor	Single organization	Combination	Specific community
Hosting	Shared resources	Dedicated infrastructure	Integrated environments	Shared infrastructure
Control	Least control	Greater control	Variable control	Tighter control
Cost	Cost-effective	Can be expensive	Optimized workload placement	Cost sharing
Best For	Economies of scale	Security, compliance	Workload optimization	Shared requirements

Fig. 10.3 Comparative analysis of cloud deployment models

10.3 Virtualisation in Cloud Computing

Virtualisation technology serves as the fundamental enabler of cloud computing, providing the abstraction layers and resource management capabilities that make elastic, scalable, and efficient cloud services possible. Understanding virtualisation concepts, implementation approaches, and management tools is essential for effective cloud computing deployment and management [5].

10.3.1 Basics of Virtualisation

Virtualisation creates logical abstractions of physical computing resources, enabling multiple virtual instances to share underlying hardware while maintaining isolation and security between different workloads. This fundamental capability enables the resource pooling, elasticity, and multi-tenancy characteristics that define cloud computing environments [6].

The core concept of virtualisation involves inserting a virtualisation layer, typically called a hypervisor or virtual machine monitor, between physical hardware and operating systems. This layer manages physical resources and presents virtualised resources to guest operating systems, creating the illusion of dedicated hardware while enabling efficient sharing of actual resources.

Virtual machines created through virtualisation technology behave like independent computers with their own operating systems, applications, and data, despite sharing physical infrastructure with other virtual machines. This isolation ensures that problems in one virtual machine do not affect others, while resource management capabilities enable dynamic allocation of CPU, memory, storage, and network resources based on demand.

10.3.2 Types of Virtualisation

Virtualisation technology has evolved to encompass various types of resources and use cases, each addressing specific requirements for abstraction, management, and optimisation of computing infrastructure [7].

Server Virtualisation represents the most common and widely deployed form of virtualisation, enabling multiple virtual machines to run on a single physical server. Type 1 hypervisors, also known as bare-metal hypervisors, run directly on physical hardware without requiring a host operating system. Citrix XenServer, Microsoft Hyper-V Server, VMware vSphere/ESXi, and KVM (Kernel-based Virtual Machine) are a few examples.

Although type 2 hypervisors are simpler to install and administer because they operate as applications on top of host operating systems, the overhead of the host OS layer may make them less effective. Parallels Desktop, Oracle VirtualBox, and VMware Workstation are a few examples.

In order to establish logical networks that function independently of the underlying physical network infrastructure, network virtualisation abstracts network resources. A thorough method of network virtualisation that divides network control plane and data plane forwarding operations is software-defined networking, or SDN.

Storage virtualisation creates logical storage pools that can be dynamically distributed to virtual machines and applications by abstracting away physical storage resources. This method facilitates advanced features like snapshots, cloning, and replication while also increasing storage utilisation and streamlining administration.

By hosting desktop operating systems in data centre virtual machines, desktop and application virtualisation allows users to access their desktop environments from any device. While application virtualisation isolates applications from the underlying operating system, virtual desktop infrastructure (VDI) implementations offer separate virtual machines for every user.

10.3.3 Virtualisation Tools and Hypervisors

Performance, feature availability, and management complexity in cloud environments are all greatly impacted by the choice and use of suitable virtualisation tools and hypervisors [7].

The most sophisticated and feature-rich enterprise virtualisation platform is VMware vSphere, which offers extensive ecosystem support, sophisticated features, and management tools. Key vSphere features include vMotion for live migration of virtual machines, Distributed Resource Scheduler (DRS) for automatic load balancing, and High Availability (HA) for automatic restart of failed VMs.

KVM (Kernel-based Virtual Machine) provides open-source virtualisation capabilities that are integrated directly into the Linux kernel. KVM offers excellent performance and scalability while supporting a wide range of guest operating systems and hardware platforms. Management tools include libvirt for virtualisation API abstraction and oVirt for comprehensive data centre virtualisation management.

Xen Hypervisor pioneered paravirtualisation techniques and continues to provide high-performance virtualisation capabilities for demanding workloads. Amazon Web Services uses customised Xen hypervisors for many of its EC2 instance types, demonstrating the scalability and reliability of Xen technology.

Microsoft Hyper-V provides integrated virtualisation capabilities within Windows Server environments, offering seamless integration with Microsoft management tools and applications. Hyper-V features include live migration, failover clustering, and replica-based disaster recovery.

10.3.4 Containers and Orchestration

Container technology has emerged as a complementary and sometimes alternative approach to traditional virtualisation, offering lightweight application isolation and efficient resource utilisation [8].

Docker revolutionised container adoption by providing user-friendly tools for creating, distributing, and running containerised applications. Docker containers encapsulate applications and their dependencies in portable units that can run consistently across different environments. The lightweight nature of containers enables higher-density deployments compared to virtual machines, as containers share the host operating system kernel.

Kubernetes has become the dominant platform for container orchestration, providing automated deployment, scaling, and management capabilities for containerised applications. Key Kubernetes concepts include pods as the basic deployment units, services for network access and load balancing, and deployments for managing the application lifecycle.

OpenShift offers an enterprise Kubernetes distribution, complemented by additional developer and operational tools, enhanced security features, and commercial support from Red Hat. OpenShift includes integrated CI/CD pipelines, developer workflows, and administrative tools that simplify Kubernetes adoption in enterprise environments.

10.3.5 Virtualisation Versus Containerisation

The choice between traditional virtualisation and containerisation depends on specific requirements, including isolation needs, performance characteristics, management preferences, and application architectures [9].

Virtual machines provide strong isolation through complete guest operating systems and hypervisor enforcement of resource boundaries, making VMs suitable for multi-tenant environments with strict security requirements or legacy applications with specific OS dependencies.

Containers provide lightweight isolation through shared kernel mechanisms and namespace separation, enabling higher density deployments and faster startup times. Container isolation is generally sufficient for applications that trust each other and do not require strict security boundaries.

Hybrid solutions enable organisations to enjoy the benefits of both the virtual machines (VMs) and containers. As an example, containerised applications can be executed within VM security boundaries or containers can be developed as then deployed to VM security production environments.

10.4 Cloud Service Providers and Platforms

Most of the market in cloud computing is dominated by a few huge providers who are offering comprehensive suites of services, including platforms, applications, and infrastructure. It is important to know the positioning, strengths, and capabilities of these providers to make educated decisions on cloud adoption and multi-cloud strategies [10].

10.4.1 Amazon Web Services (AWS)

Amazon Web Services pioneered the market of the modern cloud computing and has preserved its leading position due to continuous innovation, wide range of services and construction of the global infrastructure. The list of 200 plus full-fledged services that AWS offers only includes compute, storage, databases, analytics, machine learning, and specialised industry solutions.

Examples of core AWS compute services include Elastic Compute Cloud (EC2) to create virtual servers, Lambda to create serverless computing, Elastic Container Service (ECS) and Elastic Kubernetes Service (EKS) to create container orchestrators, and Batch to create large-scale parallel processing.

The AWS storage services include Simple Storage Service (S3) object storage, Elastic Block Store (EBS) persistent block storage and Elastic File System (EFS) shared file storage. Examples of database services include Relational Database

Service (RDS) of the managed SQL database, DynamoDB of NoSQL applications, and Redshift of data warehousing.

The AWS worldwide system provides disaster recovery service and low-latency access all around the world, including multiple availability segments in diverse geographic regions.

10.4.2 Microsoft Azure

The Microsoft Azure is an all rounded cloud service provider with a high level of integration to the Microsoft technologies through leveraging the large enterprise connections of Microsoft and its software ecosystem. Azure provides hybrid cloud features that enable seamless integration between cloud services and Microsoft environments on-premises.

Azure compute services include Functions for serverless computing, App Service for PaaS applications, Virtual Machines for IaaS workloads, and Container Instances for containerised applications. Azure Active Directory integrates with on-premises Active Directory deployments by offering identity and access management services.

Office 365 integration enables workflows for productivity and collaboration across desktop programs, cloud services, and mobile devices. Azure offers specialised services for the government, financial services, healthcare, and other regulated sectors.

10.4.3 Google Cloud Platform (GCP)

Google Cloud Platform offers cutting-edge cloud services with robust performance characteristics and sophisticated analytics capabilities by utilising Google's experience in large-scale distributed systems, machine learning, and data analytics.

Compute Engine for virtual machines, App Engine for hosting applications, Cloud Functions for serverless computing, and Google Kubernetes Engine (GKE) for container orchestration are examples of GCP compute services.

Particularly strong points of GCP are its data and analytics offerings, which include BigQuery for analytics and data warehousing, Cloud Dataflow for batch and stream processing, and a range of machine learning services. Pre-trained APIs for common AI tasks and AutoML for custom model development are examples of machine learning and artificial intelligence services.

10.4.4 IBM Cloud and Oracle Cloud

IBM Cloud focuses on meeting the needs of enterprises and hybrid cloud environments, leveraging IBM's expertise in enterprise computing, artificial intelligence, and industry-specific solutions. Watson AI services provide industry-specific artificial intelligence capabilities, while Red Hat integration provides comprehensive hybrid cloud and container capabilities through OpenShift.

Oracle Cloud Infrastructure emphasises high-performance computing, database services, and enterprise applications. Autonomous database services represent Oracle's differentiation in the cloud market, providing self-managing database capabilities that reduce administrative overhead while improving performance and security. Figure 10.4 compares major cloud service providers—Amazon Web Services (AWS), Microsoft Azure, Google Cloud Platform (GCP), IBM Cloud, and Oracle Cloud—highlighting their unique offerings. In August 2025, AWS will have more than 200 fully functional services, Azure will have hybrid cloud capabilities, GCP will leverage Google's large-scale machine learning expertise, IBM Cloud will concentrate on enterprise hybrid computing, and Oracle Cloud will prioritise high-performance enterprise applications to meet a variety of industry demands.

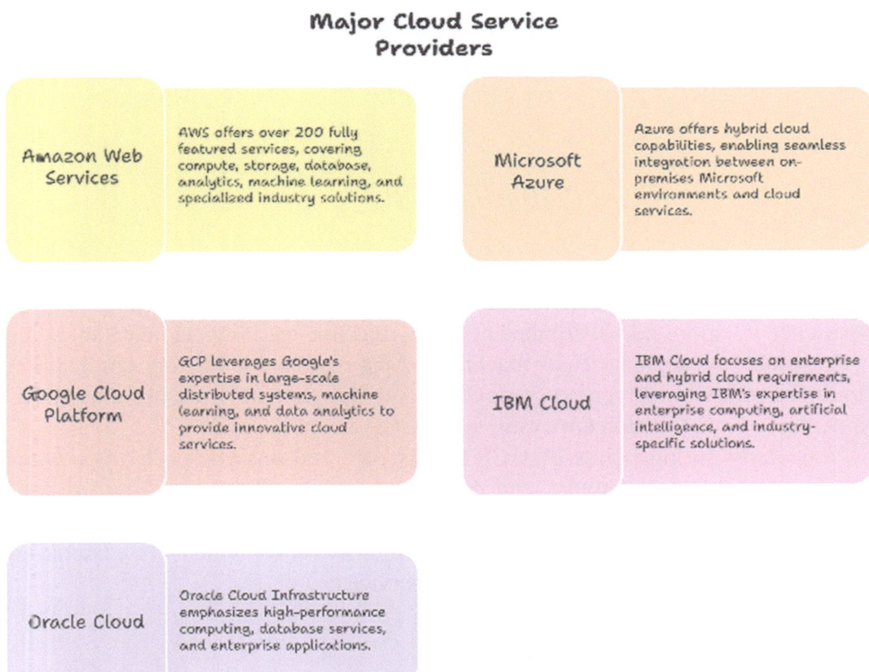

Fig. 10.4 Comparative overview of major cloud service providers

10.4.5 Comparative Analysis of Providers

Numerous factors, including technical requirements, cost considerations, geographic presence, compliance requirements, and strategic technology preferences, influence the selection of suitable cloud providers. Every significant supplier has unique benefits and positioning that cater to various market niches and use cases.

AWS remains the leader on the basis of the total market share, although other providers are more effective in specific markets or types of services. The cost optimisation plans by providers and their pricing models are highly different, which requires a critical analysis of the overall cost of ownership.

Other providers are focused on specific strengths but AWS provides the biggest service catalogue. Services differ by far in terms of scope and depth. In organisations that have specific data residency needs or regulatory requirements, the choice of a provider is determined by the presence and compliance.

10.5 Emerging Trends in Cloud Computing

The ongoing rapid change in the cloud computing landscape is driven by rapid technological changes, evolving business demands, and new applications that are pushing existing cloud computing methodologies to their limits [11–13].

10.5.1 Serverless Computing

Serverless computing also known as Function-as-a-Service (FaaS) is a fundamental shift in how applications are designed, deployed and managed in cloud environments. The paradigm provides pay as you execute models, auto-scaling, and enhanced high availability and offloads server management responsibilities on the developers.

AWS Lambda invented serverless computing, which continues to dominate the market due to its extensive connection to other AWS services and the ability to support many programming languages. The serverless computing platform of Microsoft, Azure Functions, is highly integrated with Azure services and Microsoft development tools. Google Cloud Functions focuses on easy and integration with Google Cloud services and focuses on event-driven, lightweight functions.

Unlike traditional applications that are mono-architectural, the serverless computing model requires alternative development approaches because more recent architecture patterns are being introduced: microservices decomposition, event-driven architecture, and stateless application design.

10.5.2 Edge and Fog Computing for IoT Integration

The growth of edge and fog computing architectures, which can bring cloud computing services nearer to data sources and end-users, has been driven by the increase in Internet of Things (IoT) devices and applications that require real-time processing.

Computational resources are positioned at the network edge by edge computing, usually near sensors and IoT devices. This approach enables real-time processing of sensor data, immediate response to critical events, and reduction of network traffic to centralised cloud resources.

Major cloud providers have developed edge computing platforms that extend their cloud services to the edge of their networks. AWS IoT Greengrass enables local processing of IoT data using Lambda functions, while Microsoft Azure IoT Edge provides containerised workloads that can run machine learning models and process streaming data at edge locations.

10.5.3 Multi-Cloud and Hybrid Cloud Strategies

Organisations increasingly adopt multi-cloud and hybrid cloud strategies to avoid vendor lock-in, optimise costs, leverage best-of-breed services, and meet diverse regulatory or geographic requirements.

Multi-cloud strategies utilise services from multiple cloud providers simultaneously, enabling organisations to select optimal services for specific workloads while maintaining negotiating leverage with providers. Standardised management tools, uniform security regulations, and portable application architectures are necessary for the successful deployment of multi-cloud strategies.

Cloud services and on-premises infrastructure are combined in hybrid cloud architectures to create integrated environments that strike a balance between cost, control, and compliance. Cloud management platforms provide unified interfaces for resource management across various cloud and on-premises environments.

10.5.4 Green and Sustainable Cloud Computing

Due to regulatory requirements, corporate sustainability commitments, and increased awareness of the environmental impact of technology, environmental sustainability has emerged as a crucial factor in cloud computing.

With significant investments in renewable energy projects and carbon offset programs, major cloud providers have committed to achieving carbon neutrality and increasing their use of renewable energy. In order to minimise energy consumption, energy-efficient data centre design integrates sophisticated cooling systems, optimised server configurations, and intelligent workload management.

The goal of green software development techniques is to create effective applications that use less energy and resources. This includes designing databases efficiently, minimising data transfer, and optimising algorithms.

10.5.5 AI and Cloud Synergy

Cloud computing and artificial intelligence combine to provide strong platforms for the large-scale development, training, and implementation of AI applications. Cloud vendors deliver the computing power needed to run large-scale AI deployments, and also offer end-to-end AI services that make advanced machine learning technology more democratic.

Service platforms provide pre-trained models and APIs, AI-as-a-Service provides the most simple AI tasks such as image recognition, natural language processing, and predictive analytics. To train complex machine learning models, cloud-based artificial intelligence development platforms provide specialised hardware, including GPUs and TPUs. They also have automated AI application deployment pipelines and data preparation and model training managed services.

10.6 Security and Privacy in Cloud Computing

The security and privacy issues of cloud computing involve specific challenges caused by distributed architectures, multi-use of infrastructure, and sharing security responsibilities with external vendors [14, 15].

10.6.1 Cloud Security Challenges

Some of the cloud environments are provided by multiple service providers and tenants. This creates issues regarding cross-tenant attacks, isolation of the cloud infrastructure and data. Customers are still in charge of data security, access control, and application-level security measures, while providers usually handle infrastructure security.

10.6.2 Security Mechanisms

Identity and Access Management (IAM) systems provide centralised control over user authentication, authorisation, and access policies across cloud environments. Multi-factor authentication (MFA) requirements enhance security by requiring multiple independent authentication factors before granting access to cloud resources [16].

Encryption and Secure Data Storage protect data confidentiality both in transit and at rest, providing protection against unauthorised access even if the underlying storage or network infrastructure is compromised. Key management systems provide secure generation, storage, rotation, and destruction of encryption keys.

Intrusion Detection and Prevention systems monitor network traffic and individual systems for suspicious activities and known attack patterns. Security information and event management (SIEM) systems aggregate security logs and events from multiple sources to provide comprehensive security monitoring and incident response capabilities.

10.6.3 Compliance and Governance

GDPR compliance in cloud environments requires an understanding of data processing activities, implementation of privacy-by-design principles, and establishment of appropriate data processing agreements with cloud providers. HIPAA compliance for healthcare organisations requires business associate agreements with cloud providers and implementation of appropriate safeguards for protected health information [17].

10.6.4 Best Practices for Secure Cloud Adoption

Cloud security best practices encompass planning, implementation, and ongoing management activities that address the unique challenges of cloud computing environments. Security assessments and due diligence should evaluate a cloud provider's security capabilities before selecting providers and services.

Architectural design should incorporate security principles, including defence in depth, least privilege access, network segmentation, and encryption throughout the system design process. Configuration management ensures that cloud resources are deployed with appropriate security settings and maintained in secure configurations throughout their lifecycle.

10.7 Real-World Applications and Case Studies

Cloud computing implementations across various industries demonstrate the transformative potential of cloud technologies while providing insights into practical challenges, success factors, and lessons learned from large-scale deployments [18–20].

10.7.1 Healthcare

Healthcare organisations have adopted cloud computing to enhance patient care, reduce costs, and facilitate innovative care delivery models while meeting stringent regulatory and security requirements. The migration of Electronic Health Records (EHR) to cloud platforms enables healthcare organisations to reduce IT infrastructure costs while improving system reliability and disaster recovery capabilities.

Telemedicine platforms leveraged cloud computing capabilities during the COVID-19 pandemic to rapidly scale telehealth services and maintain continuity of patient care. Cloud-based medical imaging solutions offer scalable storage and processing capabilities for large medical image files, while also enabling AI-powered diagnostic assistance.

10.7.2 Government and E-Governance Initiatives

The government's adoption of cloud computing enables improved citizen services, cost reduction, and enhanced collaboration, while addressing unique security, compliance, and sovereignty requirements. Digital India initiatives utilise cloud computing to deliver digital services to citizens at scale, while reducing costs and enhancing accessibility.

US GovCloud implementations by major cloud providers provide specialised cloud services that meet government security, compliance, and sovereignty requirements while enabling agencies to leverage commercial cloud capabilities.

10.7.3 Industry Case Study: Netflix on AWS

Netflix's migration to AWS represents one of the most significant and well-documented cloud transformation initiatives, demonstrating the potential for cloud computing to enable business transformation and global scale. The migration enabled Netflix to scale globally while maintaining high availability and performance for streaming video services.

Netflix's microservices architecture leverages AWS services to create a resilient, scalable platform, while its approach to chaos engineering demonstrates proactive resilience testing that identifies and addresses potential failures before they impact customers.

10.7.4 Smart Cities and IoT-Cloud Integration

Smart city initiatives demonstrate the integration of IoT technologies with cloud computing platforms to create comprehensive urban management systems. Barcelona Smart City implementation integrates thousands of IoT sensors with cloud-based analytics platforms to optimise city services, while Singapore Smart Nation initiatives leverage comprehensive IoT and cloud integration for data-driven urban planning.

10.7.5 Education and Remote Learning Platforms

The COVID-19 post-era has seen a significant increase in online tool-based lecture delivery. These online tools are based on cloud models, as the software is deployed in the cloud and the data is analysed in the cloud as well.

10.8 Challenges and Future Research Directions

The continued evolution of cloud computing presents significant challenges that necessitate innovative solutions and research advancements.

10.8.1 Scalability and Cost Optimisation Issues

While cloud computing offers theoretically unlimited scalability, practical implementations face constraints related to network bandwidth, data locality, and cost optimisation. Auto-scaling algorithms must balance responsiveness to demand changes with cost optimisation, while the complexity of cloud pricing models makes cost optimisation challenging.

10.8.2 Interoperability and Standardisation Gaps

The lack of comprehensive standards for cloud interoperability creates vendor lock-in risks and complicates the implementation of multi-cloud and hybrid cloud environments. Container standards have improved application portability, but differences in networking, storage, and security implementations still create compatibility challenges.

10.8.3 Sustainability and Carbon Footprint Reduction

The environmental impact of cloud computing continues to grow with increasing adoption and data centre expansion. Carbon footprint measurement and reporting require standardised methodologies, while green computing research focuses on developing more energy-efficient algorithms and integrating renewable energy.

10.8.4 AI-Driven Cloud Resource Allocation and Optimisation

The integration of artificial intelligence and machine learning into cloud resource management promises to improve efficiency through intelligent automation. Predictive resource management utilises machine learning algorithms to forecast demand patterns and automatically adjust resource allocations.

10.8.5 Quantum Computing and Its Impact on Cloud

The emergence of quantum computing technologies will significantly impact cloud computing through new service offerings and fundamental changes to security and computational capabilities. Quantum cloud services are emerging, while post-quantum cryptography will become increasingly essential as quantum computers continue to develop.

10.9 Conclusion

Summary of Key Takeaways

Cloud computing has fundamentally transformed the technology landscape by democratising access to enterprise-grade computing resources, enabling new business models, application architectures, and operational approaches. The maturation of virtualisation technologies and the development of comprehensive service ecosystems have created a robust foundation for continued growth and innovation in cloud computing.

Security and compliance capabilities have matured significantly, addressing many early concerns about cloud computing adoption in regulated industries. The emergence of edge computing, serverless architectures, and AI-driven optimisation represents continued innovation that addresses evolving requirements for performance, efficiency, and functionality.

Role of Cloud Computing in Digital Transformation

Cloud computing serves as a fundamental enabler of digital transformation by providing the technological foundation for modern applications, data analytics, and business process optimisation. The democratisation of advanced technologies through cloud services enables smaller organisations to access capabilities previously available only to large enterprises.

Agile development practices and continuous deployment models are enabled by cloud platforms that provide consistent development, testing, and production environments with automated provisioning and management capabilities.

Outlook for the Future

The future of cloud computing will be characterised by continued innovation in artificial intelligence integration, edge computing expansion, sustainability improvements, and quantum computing integration. The integration of AI and machine learning into cloud platforms will enable increasingly intelligent and autonomous systems.

Edge computing integration will extend cloud computing capabilities to support real-time applications and IoT deployments, while sustainability considerations will drive continued innovation in energy efficiency and the integration of renewable energy. The successful future of cloud computing depends on addressing current challenges while continuing to innovate and expand capabilities.

References

1. Upreti K, Jain R, Kumar R, Goyal K, Gupta K. The evolution of cloud computing: a study of aspirational technologies and practical achievements. In: 2025 IEEE 14th international conference on communication systems and network technologies (CSNT). IEEE; 2025. p. 898–903.
2. Jain J, Dixit A. Cloud computing and virtualization: a symbiotic revolution in modern IT infrastructure. In: Web 3.0. CRC Press; 2024. p. 24–38.
3. Stamenkov G. Cloud service models, business continuity and disaster recovery plans, and responsibilities. Int J Organ Anal. 2025;33(3):437–51.
4. Saha B. A comparative study of cloud deployment strategies for GANs: public versus private clouds. Private Clouds; 2025.
5. Venkatesha GG, Sharma R, Marella V, Acharya V, Sivaraj KP, Shah S. Quantifying performance trade-offs in network virtualization for cloud computing environments. In: 2025 International conference on pervasive computational technologies (ICPCT). IEEE; 2025. p. 130–134.
6. Dong X, Xie Y. Research on cloud computing network security mechanism and optimization in university education management informatization based on OpenFlow. Syst Soft Comput. 2025;2: 200225.
7. Sharma S, Kumar DL, Vijay EA. Introduction to virtualization security and cloud security. In: Machine learning hybridization and optimization for intelligent applications. CRC Press; 2025. p. 108–126.
8. Mahavaishnavi V, Saminathan R, Prithviraj R. Secure container orchestration: a framework for detecting and mitigating orchestrator-level vulnerabilities. Multimed Tools Appl. 2025;84(17):18351–71.
9. Melo P, da Silva GJ, Vanderlei AB, Beserra D, Araujo J. Performance evaluation of containerized virtualization platforms. SN Comput Sci. 2025;6(7):771.
10. Borra P. An overview of cloud computing and leading cloud service providers. Int J Comput Eng Technol (IJCET). 2024;15:122–33.
11. Dritsas E, Trigka M. A survey on the applications of cloud computing in the industrial internet of things. Big Data Cogn Comput. 2025;9(2):44.
12. Younus M, Purnomo EP, Nurmandi A, Mutiarin D, Manaf HA, Mumtaz F, Khairunnisa T. Analyzing the trend of government support for cloud computing usage in e-government architecture. J Cloud Comput. 2025;14(1):14.
13. Amudha G, Gopika P, Dinesh MG. Cloud computing: transformations, opportunities, and challenges. In: 2025 3rd International conference on self sustainable artificial intelligence systems (ICSSAS). IEEE; 2025. p. 897–903.
14. Salama R, Al-Turjman F. Mobile cloud computing and the internet of things security and privacy. In: Edible electronics for smart technology solutions. IGI Global; 2025. p. 333–350.
15. Albogami NN. Intelligent deep federated learning model for enhancing security in internet of things enabled edge computing environment. Sci Rep. 2025;15(1):4041.
16. Szymoniak S, Piątkowski J, Kurkowski M. Defense and security mechanisms in the internet of things: a review. Appl Sci. 2025;15(2):2076–3417.
17. Li Y. Optimising corporate governance with internet of things and artificial intelligence: a data-driven framework for legal systems. Int J Inf Commun Technol. 2025;26(13):81–100.
18. Hoang ML. A comprehensive review of machine learning, and deep learning in wearable IoT devices. IEEE Access; 2025.
19. ur Rehman A, Alblushi IG, Zia MF, Khalid HM, Inayat U, Benbouzid M, Muyeen SM, Hussain GA. A solar-powered multi-functional portable charging device (SPMFPCD) with internet-of-things (IoT)-based real-time monitoring—an innovative scheme towards energy access and management. Green Technol Sustain. 2025;3(1):100134.
20. Tripathi P, Sasidhar K, Mistry H, Shah V. Scheduling computing tasks on smartphones: comparative case studies of metaheuristic algorithms on real world applications. SN Comput Sci. 2025;6(4):298.

Chapter 11
Human–Computer Interaction in the Age of Intelligent Systems

11.1 Introduction

11.1.1 Evolution of Human–Computer Interaction

The evolution of Human–Computer interaction as an engineering field of study has seen a radical shift in its early days where maximisation of the human performance with computers was witnessed to a multi-faceted field that influences the way billions of individuals interact with technology on a daily basis [1]. The change in the form of the field itself (the personal computers, mobile technology, and even the ambient intelligent technology) denotes a cyclic shift in the fundamentals of computing in which what can be born as a room-sized mainframe can transform into a cellular mobile device and lastly a seamlessly accessible ambient presence so likely in our environment that it becomes institutionalised [2].

The early phase of HCI, from the 1960s to the 1980s, focused on the efficiency of engineering and cognitive ergonomics. The concepts applied included user feedback, error prevention, and interface consistency. The second movement expanded the field's coverage by taking into account not only social and contextual aspects but also utilising ethnography alongside the working systems. The third wave currently underway is marked by the marriage of artificial intelligence and machine learning, as well as ubiquitous computing technologies that allow customisation, tailoring, and situational interaction.

Modern HCI is used to include various modalities of interaction, such as voice interfaces, gesture recognition devices, brain–computer interfaces, and a virtual environment that surrounds human interactions. This growth requires provision of new theoretical perspectives, design methodologies, and assessment strategies taking into consideration the complexity of intelligent systems and how they can be incorporated into human social and cultural environments [3]. Figure 11.1 gives an overview graphic about development of Human–Computer interaction.

T. H. Sardar, *The Spectrum of Computer Science: Emerging Technologies and Trends*, Studies in Smart Technologies, https://doi.org/10.1007/978-3-032-19632-3_11

Fig. 11.1 Evolution of human–computer interaction

11.1.2 Fundamental Design Principles

The User-Centered Design (UCD) is still the HCI practice backbone, which focuses on the development processes in iterations, to engage the user in the process [4]. UCD methods start with rigorous user research involving interviews, observations, and contextual inquiries of who, what the users want to accomplish and what they cannot do so they can know the desired outcomes, tasks, and constraints of the user. These are used to inform persona development, scenario creation, and journey mapping which inform design.

This principle of affordances which was first proposed by Gibson but modified by Norman describes the actions that interfaces can expose the user to [5]. The element of efficient interface design guarantees the correspondence between the perceived and the real system functionality with the well-chosen signifiers of visual, auditory, or tactile feedback providing the information about the opportunities of interaction.

System state and action feedback mechanisms make the users informed about the outcomes of the actions and allow controlling the system and gaining confidence. The principles of consistency ease the cognitive load by trying to make the same interface components to behave in the same way, in different contexts. Defense and recover mechanisms take into consideration that errors are always undesirable, and the problem is how to minimise their frequency and effects.

11.2 Human Factors in Interaction Design

11.2.1 Cognitive Considerations

Cognitive Load Theory offers a paradigm to comprehend the case of information processing constraint on the user performance using interactive systems [6]. The theory differentiates intrinsic load (complexity of a task), extraneous load (poor design) and the germane load (constructive learning). Anything that can reduce extraneous cognitive load is the effective interface design with the clear visual hierarchy, spread information logically and progressive disclosure approaches.

Mental models are a form of inward knowledge of the system operation of users who acquire it through the course of experience and edification. The interface design should provide proper mental development of the mental models with clear conceptual framework, uniform patterns of behaviours, and easy indication of system operations. This is especially a challenge to intelligent systems where the algorithmic processing might not be in line with the intuitive expectations of the users.

The constraints of the working memory have great implication on the articulateness of a user to process and accomplish complicated tasks. Mainly, research has1 pointed out that working memory has the capacity to store $7 + 2$ pieces of information at once [7]. These limitations need to be considered in the interface design with the help of external memory, e.g., the use of breadcrumbs, recently visited objects, and persistent state representation.

11.2.2 User Experience Dimensions

The usability involves effectiveness (accuracy in completing the task), efficiency (resource demand), and satisfaction (train of thoughts of the end user) [8]. Completion rates, error rates, and time-on-task are some traditional measurements but intelligent systems need even more intelligence metrics like trust in algorithm recommendations and long-term adoption behaviour.

Accessibility is a rule that makes systems accessible to individuals of different capabilities, based on guidelines, including the Web Content Accessibility Guidelines (WCAG). Universal Design principles foresee the concept that inherently accessible systems can be useful to all users instead of special adaptations.

The concept of emotional design acknowledges the fact that feelings towards the product in terms of emotional response are important aspects that drive perceived usability and the experience which one gets using the product. Negative emotions put more emphasis on perceived system flaws whereas positive emotional reactions cause one to tolerate lesser usability issues. The design is effective in regard to immediate emotional responses and the more enduring emotional relations that the people will have with the systems in the long term.

11.3 Emerging Interaction Paradigms

11.3.1 Immersive Technologies

VR produces completely artificial worlds that the user is enclosed in with computer generated experience of every sense. VR systems should have the ability to allow them to feel that they are present in a virtual space [9], which is important and determines their usefulness. Presence represents a result of the sensory faithfulness, naturalness of interaction, and narrative involvement.

Modern VR applications include entertainment applications, educational applications, training applications, and therapeutic applications. Training programs use the capabilities of VR to imitate real life training of dangerous conditions, and educational programs give students an experience of learning that they cannot have in the real world. VR interface design involves knowledge of principles of interaction in space that cannot be applied to a non 2D interface.

Augmented Reality (AR) overlays digital information onto the physical world, creating hybrid environments that combine real and virtual elements. AR applications range from simple information overlays to complex spatial manipulation systems. Mobile AR uses smartphone sensors for surface detection and object recognition, while head-mounted displays enable hands-free interaction.

AR interface design requires solving challenges in accurate tracking, reliable object recognition, and seamless integration of virtual content with real-world lighting and perspective. Privacy and social acceptance represent additional challenges, particularly for head-mounted displays that may record environments without clear indication to bystanders.

11.3.2 Conversational Interfaces

Natural Language Processing (NLP) enables computers to understand and generate human language for natural conversation [10]. Contemporary systems leverage transformer architectures and large language models that demonstrate sophisticated understanding of linguistic nuances and conversational patterns.

Effective dialogue systems require understanding of syntax, semantics, pragmatics, and discourse-level patterns. Intent recognition identifies users' underlying goals, while entity extraction identifies specific information relevant to task completion. Dialogue state tracking maintains conversation history and context across multiple turns [11].

Trust development in conversational interfaces depends on perceived system competence, benevolence, and integrity. Transparency mechanisms help users understand system capabilities and limitations, though excessive explanation can disrupt natural conversation flow. Cultural and linguistic diversity presents challenges for global deployment, requiring adaptation to different communication styles and social

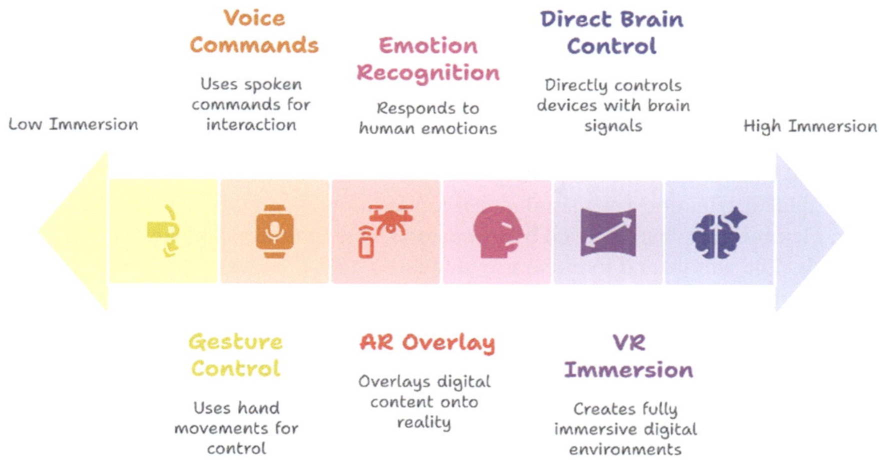

Fig. 11.2 Emerging interaction paradigms ranked by level of immersion

norms. Figure 11.2 provides different emerging interaction paradigms ranked by the level of immersion.

11.3.3 Brain–Computer Interfaces

Brain–Computer Interfaces (BCIs) create direct communication pathways between the nervous system and computing systems, bypassing traditional sensory and motor channels [12]. Contemporary systems primarily use electroencephalography (EEG) to measure brain electrical activity, though research continues into other approaches including functional near-infrared spectroscopy and implanted arrays.

Neuroadaptive interfaces adjust behaviour based on real-time measurements of cognitive and emotional states, detecting mental workload, attention, and frustration. Applications include educational systems that adapt difficulty based on cognitive load and assistive technologies for individuals with severe motor impairments.

BCI development faces technical challenges including signal noise, individual differences, and extensive calibration requirements. The user experience differs fundamentally from traditional interfaces because users must learn to modulate brain activity rather than operate external devices.

Ethical considerations include mental privacy rights, cognitive enhancement fairness, and potential for neural manipulation. Legal and technical frameworks must address unauthorised neural surveillance while preventing misuse of enhancement technologies.

11.4 Accessibility and Inclusive Design

11.4.1 Universal Design Principles

Universal Design offers seven principles in designing inherently accessible products [13]. Equal distribution gives similar distinctions of use to all users or similar alternatives. Flexibility is characterised by its customisable interfaces and alternative input methods. The simple and intuitive use includes a high focus on the visual hierarchy and expected interaction patterns.

Perceptible information is communicated effectively through various media and with reasonable contrast to the senses, whether they possess sensory powers or not. To minimise risks (hazards), tolerance entails confirmation dialogues and undo features. Blogging offers low physical exertion, allowing use with comfort with reduced fatigue. Space and size are all taken into consideration so that it gets the right size according to various motor abilities.

11.4.2 Assistive Technologies

Screen readers can translate text and interface controls to either synthesised voice speech or Braille and allow them to be accessed by blind or unseeing DDS and patients [14]. It needs to function correctly in terms of semantic marking and keyboard friendliness. Screen reader users learn advanced navigation tactics through headings, landmarks, and a list of links that must be facilitated through interface design.

The eye-tracking systems allow users with motor impairments to control the system without using their hands, as they examine eye patterns. Design challenges include distinguishing between looking and intending to interact, calibration requirements, and accuracy limitations. Dwell-based selection and blink patterns provide activation mechanisms.

Haptic feedback systems provide tactile information through vibration, force feedback, or textural simulation. Applications include navigation aids for users with visual impairments and rehabilitation systems for motor recovery. Effective haptic design requires understanding tactile perception capabilities and pattern discrimination abilities.

11.4.3 Cognitive Accessibility

Cognitive accessibility addresses diverse processing abilities, learning differences, and attention patterns [15]. Clear, simple language using plain language principles reduces cognitive load. Consistent navigation patterns enable reliable mental models.

Error prevention and recovery mechanisms are particularly important for users with cognitive disabilities.

Time-based considerations include adequate task completion time and avoiding automatic timeouts. Emotional accessibility recognises that users may experience anxiety when interacting with complex systems. Neurodiversity considerations acknowledge cognitive differences as natural human variation rather than deficits.

11.5 Artificial Intelligence in HCI

11.5.1 AI-Enhanced Interfaces

Machine Learning for Personalization enables interfaces to adapt dynamically to individual users through behavioural data analysis [16]. Recommendation systems in platforms like Netflix demonstrate effective personalisation, though they raise questions about transparency, user autonomy, and filter bubbles.

Computer Vision for Interaction recognises and interprets visual cues including facial expressions, gestures, and gaze direction. Applications range from driver monitoring systems to AR/VR environments requiring real-time object recognition. Vision-based interaction provides natural engagement but demands robust algorithms addressing occlusion, lighting variations, and cultural differences.

Natural Language Processing has been revolutionised by large language models demonstrating emergent abilities in reasoning, creativity, and multi-turn dialogue. These models maintain extended context, adapt communication styles, and provide sophisticated responses, though they may generate plausible but incorrect information requiring careful validation. Figure 11.3 illustrates the integration of AI in Human–Computer Interaction (HCI), showcasing how AI enhances user experience through adaptive interfaces, natural language processing, and predictive analytics.

11.5.2 Benefits and Enhancements

AI enables context-aware personalisation that adjusts content and features based on location, time, preferences, and device context. Real-time adaptivity allows continuous responsiveness, exemplified by adaptive tutoring systems and healthcare monitoring platforms that modify feedback dynamically.

AI significantly enhances accessibility through speech-to-text engines, predictive typing, automated captioning, and computer vision-based navigation for visually impaired users. These technologies expand digital inclusion by providing alternative interaction modalities and reducing barriers to system access.

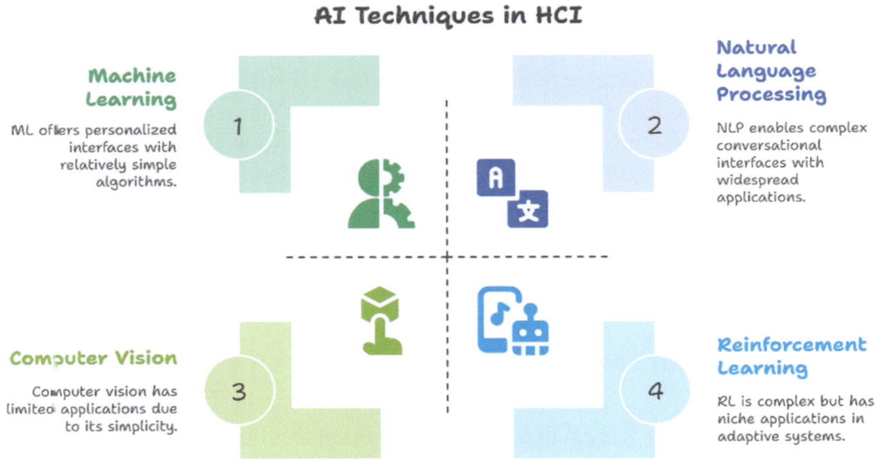

Fig. 11.3 The AI usage in HCI

11.5.3 Challenges and Risks

Privacy and Security concerns arise from the requirements of AI systems for sensitive personal data [17]. The collection of biometric and behavioural data poses the threat of unethical surveillance and potential violation. Mechanisms to ensure privacy and data security in data administration are needed to maintain users' trust.

Algorithmic bias has the ability to perpetuate discrimination in unfairly data-weighted sets or erroneous algorithms. Bias in HCI takes the form of selective performance between demographics or inequitable, closely intimate methods of HCI performance. Bias is common in how systems operate and/or distort the presentation of individual persons. To address issues related to fairness, it is essential that datasets be carefully curated and that bias be identified, with algorithmic transparency ensured.

Technical limitations include latency, scaling, and the inability to interoperate with legacy systems. It has been challenging to apply AI models across various cultural and spatial settings, and they require widespread testing and validation by their users.

The issue of Trust and Transparency becomes a concern when AI systems make decisions affecting users. Explainable AI methods allow us to understand how the system thinks, but there is a tradeoff between transparency and usability. Users must have light knowledge that is not too complex.

11.6 Future Directions

11.6.1 Human–AI Collaboration

Future HCI will emphasise synergistic collaboration where AI augments rather than replaces human capabilities [18]. Effective collaboration requires systems that maintain human agency while leveraging AI efficiency. Design challenges include appropriate task allocation, seamless handoffs between human and AI control, and interfaces that support mixed-initiative interaction.

Explainable AI (XAI) becomes increasingly important as AI systems make more consequential decisions. Users need to understand system recommendations and actions to foster appropriate trust. Research focuses on visual explanations, justification mechanisms, and interaction-level transparency that make AI reasoning accessible to non-expert users.

Figure 11.4 illustrates a 360-degree innovation-driven ecosystem shaping the future of Human–Computer Interaction (HCI), beginning with Virtual/Augmented Reality merging physical and digital realms, advancing through the Metaverse for immersive experiences, Artificial Intelligence for personalised and automated interactions, the Internet of Things for context-aware connectivity, and Brain-Computer Interfaces for direct brain-device communication, all as of August 2025. The core theme emphasises that HCI's future hinges on balancing these innovations with ethical responsibility to ensure responsible and responsive technological advancements.

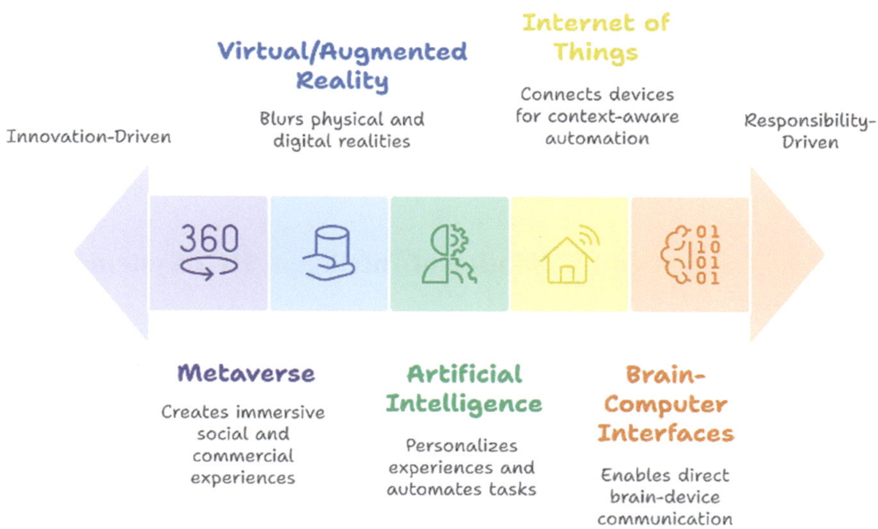

Fig. 11.4 Balancing innovation and ethics in the future of HCI

11.6.2 Sustainable and Ethical Design

Sustainability has emerged as a critical design consideration, with Green HCI emphasising low-power computation, resource-aware interaction design, and minimalism [19]. Designers must balance functionality with ecological responsibility, developing interfaces that reduce environmental impact while maintaining usability.

Responsible innovation is guided through ethical frameworks that are characterised by participatory design and by regulation. Informed consent, algorithmic transparency, user control, and fairness in delivering benefits of technology should be considered key factors. Ethics-by-design methods incorporate the use of morals throughout the life cycle of development.

This element is connected to the necessity of integrating various requirements of diverse sectors and it is therefore advised that all curricular units be synthesised to align cross-disciplinary objectives, including cross-cultural aims.

11.6.3 Cross-Disciplinary Integration

This element can be related to the need to focus on integrating diverse requirements of different sectors and thus it is recommended that all curricular units are aligned to the shared cross-disciplinary attainments such as cross-cultural aims. Combining neuroscience with psychology and AI would deliver more detailed models of human behaviour and cognition. Neuroadaptive interfaces that tap in to brain signals and psychologically inspired adaptive systems are the first steps toward quite individualised interaction frameworks.

The issue of culture gains greater significance with the globalisation in the implementation of technology. Interfaces should support different communication styles, social norms, groups, and value systems without making assumptions or stereotyping the preferences of the users.

11.7 Evaluation and Validation of Intelligent HCI Systems

Extensive Smart Human–Computer Interaction (HCI) systems should ensure the ability to provide credible, inclusive, and sustainable user interfaces in addition to the technical complexity of the system. Although usability studies and user-centred evaluation have always been part of traditional HCI, when artificial intelligence is integrated, aspects such as algorithmic transparency, fairness, and cognitive adaptability are introduced, placing new requirements on evaluation frameworks. This part describes the principles of methodology and practical policies for appraising AI-driven HCI, focusing on multifaceted metrics, sound and effective user study procedures, simulation-based benchmarking, and responsible evaluation behaviours.

11.7.1 Evaluation Metrics for AI-HCI

The intelligent interfaces should also be evaluated in relation to more standard measures of usability like task success, effectiveness, and the rate of errors. Important dimensions are the following:

- **Trust**: AI-driven systems must establish appropriate levels of trust. Automation bias can be caused by over-trust and impeded by under-trust. Measures include the rates of user reliance, metrics of trust calibration, and subjective trust ratings.
- **Transparency and Explainability**: It is expected that the users will know how and why an AI system generates specific output. Evaluation involves measuring user comprehension of explanations, perceived clarity, and the ability to predict system behaviour.
- **Fairness and Equity**: Intelligent interfaces must perform equitably across demographic and cultural groups. Measures of fairness are performance disparity measures, demographic parity, and subgroup error measures.
- **Cognitive Workload**: Working with adaptive AI can have different cognitive loads. Mental effort can be measured using tools such as NASA-TLX, EEG-based measures of workload, and response latency measures.
- **Emotional Engagement**: Intelligent systems are becoming more and more concerned with responding to user affect. Accurate recognition of emotions, affective alignment rating, and emotional satisfaction as reported by users offers insight into evaluation.

11.7.2 User Studies and Experimental Protocols

It takes methodological pluralism to provide adequate empirical validation:

- **Quantitative and Qualitative Approaches**: Surveys, controlled experiments, and log analyses provide measurable outcomes, while ethnographic observation, interviews, and think-aloud protocols capture contextual richness.
- **A/B and Multivariate Testing**: A/B and Multivariate Testing: It is possible to compare the various interface designs to evaluate a particular feature of AI-powered interfaces (e.g. explainable vs opaque recommendation interfaces). Statistical rigour reports assured conclusions.
- **Longitudinal Studies**: Short-term usability tests cannot fully capture how users build trust and adapt to intelligent systems over time. Longitudinal studies on adoption look at the long-term interaction and learning curves and changing mental models.
- **Ecological Validity**: Field experiments in naturalistic settings (e.g., smart homes, workplaces) ensure that results generalise beyond controlled lab contexts, capturing real-world complexities of AI integration.

11.7.3 Benchmarking and Simulation

The large-scale testing of AI-HCI systems is limited by the number of users and ethical concerns. These challenges are dealt with by simulation methods:

- **Synthetic Users and Agent-Based Models**: The virtual agents are used to model human behaviours in manipulated situations to enable the mass testing of the interaction patterns without exposing the users to exhaustion or loss of privacy.
- **Digital Twins of Users**: Given a personal user, computational models of their decisions may be used to potentially test such an AI–HCI interaction in an adaptive manner.
- **Simulation Environments**: VR-based/software-based environments are environments that are imitations of a real complex environment (e.g., autonomous driving, collaborative robotics) that help to carefully test HCI systems before deployment.
- **Benchmark Datasets and Reproducibility**: Public datasets which are available on genuine trusted websites, such as Kaggle.com and the UCI machine learning repository, are downloaded by many experts and experimented with. These benchmarking dataset-based works can be replicated and experimented with again, using new techniques to evaluate the quality among research groups.

11.7.4 Responsible Evaluation

Other than methodological rigour, evaluation should be ethically responsible:

- Test Group bias elimination: The changes in the samples of participants aimed at the representative will eliminate the systematic absence of minorities, elderly adults, and users with disabilities. Biased representation of evaluation samples may promote the concept of bias in the algorithm.
- Privacy and Data Protection: The evaluation protocols should meet the ethical requirements (e.g., GDPR, informed consent), especially with the gathering of biometric or affective data.
- Inclusiveness and Accessibility: Assessment mechanisms must take into consideration the needs of accessibility so that the marginalised groups can gain the same advantages of AI-based mechanisms.
- Sustainability Considerations Beyond the functional performance, energy conservation, carbon footprint, and ecology of smart interfaces should be considered in the assessment criteria, which is in line with the Green HCI principles.

11.8 Conclusion

Artificial intelligence associated with Human–Computer Interaction can be discussed as the essential paradigm shift that opens up new possibilities of natural, adaptive, and welcoming user experiences. AI technologies have the potential to provide personalised interfaces that can accommodate individual needs, context-specific systems that can meet situation requirements and accessible technologies that enhance digital accessibility among various populations of users.

But these advances also come with a lot of challenges which need close attention and active management. Overall, these are all aspects that contribute to concerns related to privacy due to the vast demands of data collection, whereas there is a risk of algorithmic bias perpetuating the social inequalities through system-level discrimination. The technical disadvantages of present AI, such as low reliability and cultural biases, require solid testing and validation systems.

Creating human-AI collaboration models that preserve user agency and control while enhancing, rather than replacing human capabilities, is key to the future of human–computer interaction (HCI). To build appropriate trust and facilitate well-informed decisions regarding system reliance, explainable AI techniques must make algorithmic decision-making transparent and intelligible to users.

To ensure that technological advancement serves the welfare of all people rather than just a select few commercial interests, sustainability and ethical considerations must be incorporated into the design process at every stage. This calls for regulatory frameworks that safeguard user rights while fostering innovative solutions and participatory design methodologies that involve diverse stakeholder communities.

The ability of AI-driven HCI to establish meaningful, equitable, and long-lasting human-technology relationships will ultimately determine its success rather than just its technical prowess. The principles of human-centred design remain crucial as we move towards a more digital future; they must be continuously adjusted to ensure that intelligent systems promote, rather than impede, human flourishing. To develop interfaces that are not only intelligent but also inclusive, humane, and aligned with core human values, technologists, social scientists, ethicists, and various user communities must collaborate across disciplinary boundaries.

References

1. Shrestha A, Poudel AS, Luitel YU, Koirala A, Humagain R, Shrestha S, Poudel P. Enhancing usability and reliability in digital payment platforms: overcoming the adoption challenges. Int J Hum-Comput Interact. 2025;41(8):4822–32.
2. Ho MT, Vuong QH. Five premises to understand human–computer interactions as AI is changing the world. AI Soc. 2025;40(2):1161–2.
3. Liu Z, Xue Y, Yang J, Lin W, Babichuk IS. Sustainable, biodegradable, flexible piezoelectric quaternary ammonium chitosan film pressure sensors for human motion detection and human-computer interaction. Sens Actuators, A. 2025;1(389): 116549.

4. Guerra A. Craft-based methodologies in human-computer interaction: exploring interdisciplinary design approaches. Multimodal Technol Interact. 2025;9(2):13.
5. Robinson B. Rethinking affordances and constraints in the platform era. In: Literacies in the platform society. Routledge;2025. pp. 85–99
6. Zou L, Zhang Z, Mavilidi M, Chen Y, Herold F, Ouwehand K, Paas F. The synergy of embodied cognition and cognitive load theory for optimized learning. Nat Hum Behav. 2025;21:1–9.
7. Sana F, Fenesi B. Working memory and instructional fit: reintroducing aptitude-treatment interaction in education research. Behav Sci. 2025;15(6):765.
8. Duan S. Systematic analysis of user perception for interface design enhancement. J Comput Sci Softw Appl. 2025; 1:5(2).
9. Ambika A, Shin H, Jain V. Immersive technologies and consumer behavior: a systematic review of two decades of research. Aust J Manag. 2025;50(1):55–79.
10. Feuerriegel S, Maarouf A, Bär D, Geissler D, Schweisthal J, Pröllochs N, Robertson CE, Rathje S, Hartmann J, Mohammad SM, Netzer O. Using natural language processing to analyse text data in behavioural science. Nat Rev Psychol. 2025;4(2):96–111.
11. Joshi A, Dabre R, Kanojia D, Li Z, Zhan H, Haffari G, Dippold D. Natural language processing for dialects of a language: a survey. ACM Comput Surv. 2025;57(6):1–37.
12. Patrick-Krueger KM, Burkhart I, Contreras-Vidal JL. The state of clinical trials of implantable brain–computer interfaces. Nat Rev Bioeng. 2025;3(1):50–67.
13. Ramírez-Saiz A, Baquero Larriva MT, Jiménez Martín D, Alonso A. Enhancing urban mobility for all: the role of universal design in supporting social inclusion for older adults and people with disabilities. Urban Sci. 2025;9(2):46.
14. Giansanti D, Pirrera A. Integrating AI and assistive technologies in healthcare: Insights from a narrative review of reviews. In: Healthcare, vol. 13, no. 5. MDPI;2025. p. 556.
15. Orío-Aparicio C, Álvarez-Couto M, Beunza-García S, Fenellós CB, Biencinto-López C, Rollón EE, Manero PV, Molina EC. Cognitive accessibility in educational settings: a systematic review. Eur J Spec Needs Educ. 2025;26:1–7.
16. Wang Y, Wang L, Siau KL. Human-centered interaction in virtual worlds: a new era of generative artificial intelligence and metaverse. Int J Hum-Comput Interact. 2025;41(2):1459–501.
17. Mooradian N, Franks PC, Srivastav A. The impact of artificial intelligence on data privacy: a risk management perspective. Rec Manag J. 2025;35(2):147–59.
18. George AS. Enhancing spatial computing, polyfunctional robotics, and neural augmentation for human-machine Synergy. PartnS Univers Innov Res Publ. 2025;3(2):61–73.
19. Laurell Thorslund M, Leifler O. Exploring assumptions about sustainability: towards a constructive framework for action in sustainable HCI. In: Proceedings of the 2025 CHI conference on human factors in computing systems;2025. pp. 1–13.

Chapter 12
Robotics and Autonomous Systems

12.1 Introduction

12.1.1 Motivation and Scope

Robotics and autonomous systems have become a pillar of the current technological progress, as it radically changes the way we conduct scientific studies, manufacture goods, and interact with the environment. From the first mechanical automata of antiquity to the most sophisticated autonomous systems based on artificial intelligence today, robotics has emerged as a key tool for increasing human capability, improving accuracy, and exploring uncharted territory [1].

Traditional manufacturing industries are no longer the only applications for modern robotics; it has spread to many other facets of life, such as healthcare delivery, scientific research, environmental monitoring, and even multidisciplinary complex problem-solving. To operate with minimal human intervention and maintain safety, dependability, and the ability to adapt to changing conditions, autonomous systems today integrate sophisticated sensing, artificial intelligence, and control systems [2].

This study thoroughly examines the background theory, technological features, and applications of robotics and autonomous systems. We examine how these technologies are transforming industries, opening up new avenues for research and innovation, and creating new ethical and societal challenges that necessitate careful consideration and proactive management.

12.1.2 Evolution of Robotics and Autonomy

Robotics' evolution can be succinctly divided into several distinct technological eras, each characterised by groundbreaking innovations and a growing number of applications. The mid-20th-century industrial revolution, which primarily involved

T. H. Sardar, *The Spectrum of Computer Science: Emerging Technologies and Trends*, Studies in Smart Technologies, https://doi.org/10.1007/978-3-032-19632-3_12

repetitive manufacturing tasks in highly structured environments, marked the first practical application of programmable robotic systems [3]. Although these early systems were revolutionary at the time, they were also extremely rigid and required a significant amount of environmental preparation.

Robots with basic environmental perception and adaptive behaviour were made possible by the development of computer-controlled systems in the 1970s and 1980s, which opened the door to a more sophisticated programming paradigm and sensor integration. During this time, machine vision systems, force sensors, and primitive versions of adaptive control algorithms became commonplace, laying the groundwork for operations closer to autonomy.

The late 20th and early twenty-first centuries witnessed a digital revolution that led to exponential growth in computational power, the miniaturisation of sensors, and the advancement of artificial intelligence. The advancements paved the way for the creation of autonomous cars, mobile robots, and systems that function in unstructured settings. Machine learning algorithms, which enable systems to learn from experience and adapt to novel situations, have also significantly improved robotics.

The convergence of several technological developments—groundbreaking materials science that enables the design of new actuators, quantum sensing with previously unheard-of accuracy, cloud computing that unlocks the potential for distributed intelligence, and artificial intelligence that facilitates sophisticated decision-making—defines the contemporary field of robotics. Systems that can collaborate with humans in challenging settings and perform intricate mental operations that were previously thought to be exclusively human domains are the outcome of this convergence [4].

12.1.3 Relevance in Contemporary Science and Industry

The robotics and autonomous systems are taking a significant role in modern science and industrial automation. Such technologies can be described as force multipliers of human capability and can be applied in the fields of research and industrial applications, which are otherwise unfeasible, too dangerous, or subject to other economic limitations [5].

Robotic systems have been utilised in scientific investigations to access previously inaccessible locations, such as the ocean depths, the poles, and the surfaces of other planets. They offer repeatable, consistent experimental methods that remove human variability and bias, which are essential in high precision areas, such as materials science, biochemistry, and physics. Autonomous laboratory systems provide continuous operation which significantly increases the discovery with high throughput experimentation and data acquisition [6, 7].

The industrial uses have also changed to more complex automation as intelligent manufacturing systems that can mass customise, optimise quality, and pre-emptively maintain themselves. Modern robotic systems contribute to sustainability goals by improving energy efficiency, reducing waste, and enabling circular economy principles through precise disassembly and material recovery [8].

Healthcare has undergone transformative impacts through the use of surgical robotics, rehabilitation systems, and assistive technologies, which enhance the quality of life for individuals with disabilities. Agricultural robotics addresses global food security challenges through precision farming, automated harvesting, and environmental monitoring that optimise resource utilisation while minimising environmental impact [9].

12.2 Foundations of Robotics

12.2.1 Core Components

Modern robotic systems integrate several fundamental components that work synergistically to enable autonomous operation and task execution. Understanding these core elements provides insight into the complexity and capability of contemporary robotic systems [10, 11].

Sensing Systems form the perceptual foundation of robotic systems, providing information about environmental conditions, system state, and task progress. Contemporary robots employ diverse sensing modalities including visual sensors (cameras, LiDAR, structured light systems), tactile sensors (pressure, texture, temperature), auditory sensors (microphones, ultrasonic transducers), chemical sensors (gas detection, pH measurement), and proprioceptive sensors (position encoders, accelerometers, gyroscopes). Advanced systems integrate multiple sensor types through sensor fusion algorithms that provide robust, comprehensive environmental understanding even under challenging conditions.

Actuation Systems convert control signals into physical motion, enabling robotic systems to manipulate their environment and achieve desired tasks. Electric motors remain the predominant actuator type, offering precise control, high efficiency, and excellent reliability characteristics. Hydraulic actuators provide high force output for heavy-duty applications, while pneumatic systems offer rapid response and inherent compliance for safe human interaction. Emerging actuator technologies include shape memory alloys, electroactive polymers, and biomimetic muscle systems that provide novel capabilities for specialised applications.

Control Architecture serves as the computational brain of robotic systems, processing sensor information, making decisions, and generating control commands. Modern controllers range from embedded microcontrollers for simple tasks to powerful computing platforms capable of real-time artificial intelligence processing. Control architectures have evolved from simple feedback loops to sophisticated hierarchical systems incorporating behaviour-based control, model predictive control, and machine learning algorithms that enable adaptive behaviour and autonomous decision-making.

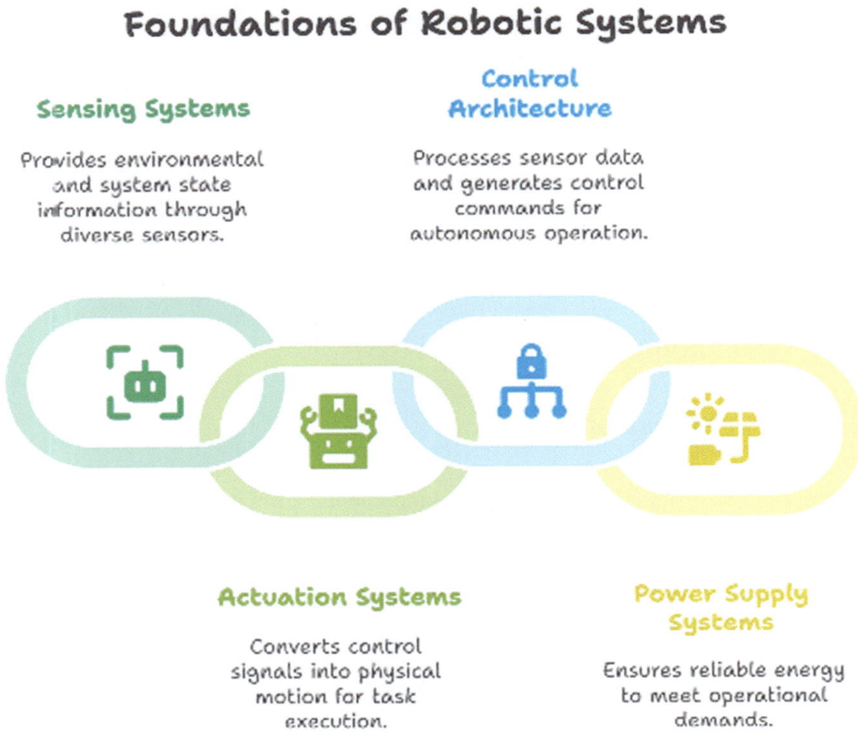

Fig. 12.1 Foundations of robotic systems

Power Supply Systems must provide reliable energy while meeting constraints of weight, size, and operational duration. Traditional battery technologies are being supplemented by fuel cells, solar panels, and energy harvesting systems that extend operational capability. Advanced power management systems optimise energy utilisation across system components, enabling extended mission duration and improved reliability.

The Fig. 12.1 illustrates the four essential components of robotic systems: sensing systems for environmental awareness, control architectures for decision-making, actuation systems for executing tasks, and power supply systems for ensuring sustained operation.

12.2.2 Classification of Robotic Systems

The diversity of robotic systems reflects the broad range of applications and operational requirements encountered across different domains. Knowing such classifications provides insight into design and trade-offs in technology [12].

Industrial Robots are the key to automated manufacturing of the modern world, which is very precise, repeatable, and fast in an organised setting. These systems are driven by a clear set of tasks and safety procedures, as well as advanced sensing features that allow for flexible production, quality screening, and cooperative working with human professionals. The current trends focus on maximising flexibility, reducing programming complexity, and enhancing safety in human-robot cooperation.

Mobile Robots are suited for navigation and use in a variety of settings, including indoor service tasks and outdoor exploration projects. They incorporate advanced navigation algorithms, environmental mapping systems and obstacle avoidance systems. Google cars are a niche segment that places special importance on safety, reliability, and regulatory compliance.

The Humanoid Robots are made to work in the human centred environment and to interact with the people in a natural manner. These systems require sophisticated balance control, dexterous manipulation capabilities, and often social interaction abilities. While challenging to develop, humanoid robots offer advantages in environments designed for human use and applications requiring natural human–robot interaction.

Collaborative Robots (Cobots) are specifically designed for safe, productive interaction with human workers. These systems have force limiting, collision detection features, and user-friendly programming interfaces that allow them to be used in shared working areas without traditional barriers to safety.

Figure 12.2 compares industrial, mobile, humanoid, and collaborative robots across key characteristics such as environments, functionalities, sensing capabilities, contemporary focus, and distinctive attributes, highlighting their specialised roles and applications.

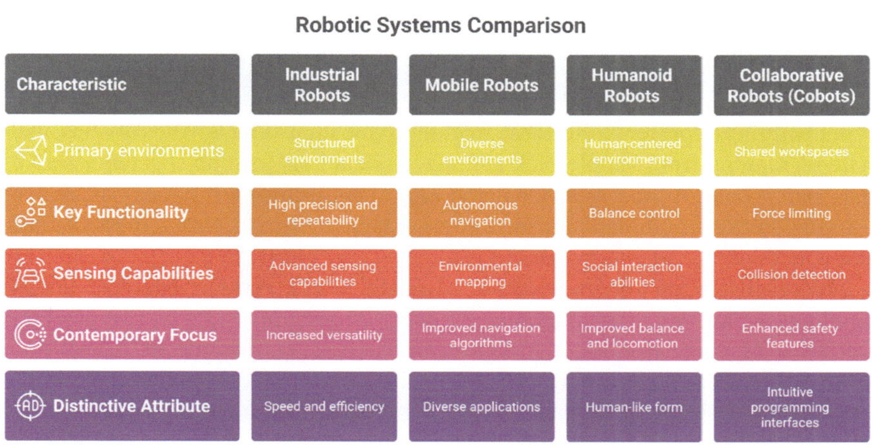

Fig. 12.2 Robotic systems comparison

12.2.3 *Autonomy Levels*

Knowledge of autonomy levels offers a way of classifying the capabilities of robots and establishing reasonable expectations of how the system should perform and human intervention [13].

Level 0: Manual Control: This is where all the actions of the robot are manually operated by using the teleoperation interfaces. This method is the most human supervised and restricts the operational range and speed, besides needing constant human attention by the operator.

Level 1: Assisted Control entails the use of automated subsystems that aids the human operators but leaves the human being in charge of making critical decisions. These include stability augmentation systems, collision avoidance and automated navigation assistance.

Level 2: Supervised Autonomy enables robotic systems to operate for an extended period, yet they require human control and intervention capabilities. Such systems are capable of doing complex mission plans as well as adapting to routine changes in their environment.

Level 3: High-autonomy systems can function independently in a broad range of conditions, making complex decisions and responding to unplanned conditions with minimal human assistance.

Level 4: Full Autonomy represents complete independent operation across all possible scenarios. While largely aspirational for complex real-world applications, this serves as a target for ongoing research and development efforts.

12.3 Applications Across Industries

12.3.1 *Manufacturing and Industrial Automation*

Manufacturing represents the most mature application domain for robotics, with systems ranging from simple pick-and-place operations to sophisticated assembly and quality control processes. Contemporary manufacturing robotics has evolved beyond traditional automation to incorporate adaptive manufacturing, mass customisation, and intelligent quality control [14].

Modern industrial robots integrate advanced vision systems enabling flexible handling of part variations, automated quality inspection, and adaptive assembly processes. Machine learning algorithms optimise process parameters in real-time, reducing waste and improving product quality. Collaborative robots are increasingly deployed alongside human workers, combining human cognitive abilities with robotic precision and endurance.

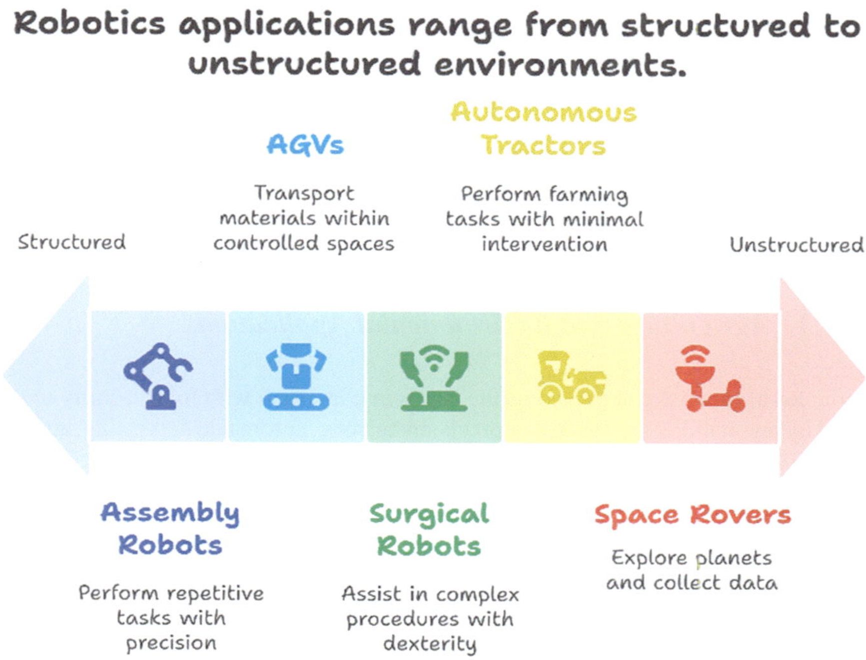

Fig. 12.3 Spectrum of robotics applications ranging from structured environments (assembly and surgical robots, AGVs) to unstructured environments (autonomous tractors, space rovers)

The Industry 4.0 paradigm has transformed manufacturing robotics through integration with digital twins, predictive maintenance systems, and supply chain optimisation. Robots now participate in complete manufacturing ecosystems, sharing information across production networks and adapting to changing demand patterns.

Figure 12.3 shows that the robotics applications across structured to unstructured environments. Structured environments, such as factories and hospitals, favour assembly and surgical robots that require precision and controlled conditions. In contrast, unstructured environments, such as farms, outer space, or disaster zones, demand autonomous tractors and space rovers capable of adapting to uncertainty.

12.3.2 Healthcare and Medical Applications

Healthcare robotics encompasses therapeutic, diagnostic, and assistive applications that directly impact patient care and quality of life. Surgical robotics has revolutionised minimally invasive procedures, providing surgeons with enhanced precision, reduced fatigue, and improved visualisation capabilities [15].

Rehabilitation robotics assists patients recovering from neurological injuries through precisely controlled exercise therapy and gait training. Such systems are predictable, quantifiable therapy but also flexible to patient improvement and ability.

Assistive robotics aids those with disabilities in their everyday living experiences, such as mobility aid and object manipulation and environmental control. These systems balance the motivation to encourage independence and those that offer safety and often use learning algorithms that adjust to specific user requirements.

12.3.3 Agriculture and Environmental Applications

Agricultural robotics can overcome the issues that interfere with food security in the world by maintaining the process of precision farming, automated harvesting, and the management of resources in a sustainable manner. With these systems, agriculture is data-driven and optimally strong in yielding crops as well as reducing environmental degradation.

GPS is guided by variable rate technology: Autonomous field robots are used to conduct precision planting, fertilising, and pesticide application tasks. On the one hand, robotised harvesting works with fragile crops, which should be carefully processed to prevent destruction, and on the other hand, labour shortage is overcome, and the quality of products is preserved [16].

The environmental surveillance systems monitor health of ecosystems using distributed sensor networks and autonomous system stations in data collection, and offer vital insight in conservation and climate change studies.

12.3.4 Scientific Research and Exploration

Introducing scientific robotics allows the study of previously inaccessible environments and scales which human scientists could only study previously under severe constraints. Robots used in space exploration work under extreme conditions in order to deliver scientific information and build infrastructure used in future explorations [17].

Robots used to explore the deep ocean at levels too deep for human exploration find new life and map ocean bottoms, as well as surveying oceanic changes.

The high-throughput experimentation, accurate handling of samples and continual use makes laboratories faster in scientific discovery allowing excessive studies to be enacted.

12.4 Robotics in Research and Innovation

12.4.1 Human–Robot Interaction (HRI)

Robotics as a research field and innovation can be incorporated in the study of engineering profits, processes of service design, and operations analysis. Research and innovation challenge into the field of Engineering profits, service design processes, and operations analysis can include robotics as a field. Human–Robot Interaction is the intersection of human research frontier which defines the feasibility of robots systems in the world of people. Proper HRI presupposes the human knowledge of human psychology, social interaction, and patterns of communication.

Natural language processing research makes robots understand and act or react to spoken instructions considering embedded context awareness and emotion recognition. Multimodal communication incorporates speech, gesture and expression in generating the experience of natural intercourse.

Research into Trust and acceptance considers the factors which control how willing human beings are to work with robotic systems. Applications that require accurate trust calibration and, consequently, an understanding of trust dynamics can greatly benefit from this understanding.

12.4.2 AI Integration and Machine Learning

Robotics has reached its height as a result of placing artificial intelligence and machine learning at its centre, creating systems that can learn, adapt, and make complex decisions in uncertain situations.

The deep learning algorithms help the robots to mill the complicated sensory data with the performance equivalent to that of the humans or any better. Convolutional neural networks are better at visual recognition, whereas recurrent networks process time sequence which is necessary in a dynamic setting.

Reinforcement learning gets robots to explore the most optimal behaviours by trial and error and possibly identifying solutions that surpass human capabilities. Transfer learning algorithms help robots transfer learning to similar issues.

12.4.3 Bio-inspired and Biomimetic Robotics

Bio-inspired robotics uses nature as the inspiration of new abilities that are beyond the conventional methods of engineering. This study will be a mixture of biological knowledge and engineering breakthrough.

Locomotion research involves the study of animal locomotion in order to formulate an efficient and versatile robotic locomotion system. The dynamics of walking robots are inspired by insights into the dynamic behaviour of quadruped animals, and the dynamics of aerial vehicles by the flight talents of birds and insects.

The flexibility of biological tissue on fish is recreated with new types of materials and the design of actuators that allow soft robotics to interact with humans with greater safety and be more adaptable to irregular surfaces.

12.4.4 Swarm Robotics

Swarm robotics is an area that involves studying how groups of simple robots can work collaboratively to reach high level objectives that surpass the life of simple robots. This study is inspired by biological swarms.

Without a centralised control, robot swarms stay in a formation, move together, and can change their configuration according to their environment. Communication protocols strike a balance between information sharing and constraints on bandwidth and energy.

Applications include environmental monitoring through distributed sensor networks, search and rescue operations, and construction projects that require the coordination of multiple robotic units.

12.5 Societal and Ethical Dimensions

Robotic systems are evolving beyond purely mechanical efficiency to engage with humans, environments, and each other in increasingly intelligent and adaptive ways. Emerging research directions highlight how advanced interaction models, AI integration, and bio-inspired designs are reshaping the next generation of robotics [18].

12.5.1 Workforce Transformation

The integration of robotic systems into the workforce represents a significant socio-economic transformation with implications extending beyond job displacement to fundamental changes in the nature of work and skill requirements.

Historical analysis reveals that technological revolutions consistently create new employment opportunities while eliminating others, but the pace and scope of robotic automation present unprecedented challenges. Unlike previous transitions affecting primarily manual labour, contemporary systems increasingly perform cognitive tasks.

Figure 12.4 outlines the ethical foundations of robotics as of August 2025, linking key principles in a chain: Impact on Employment addresses business job creation and

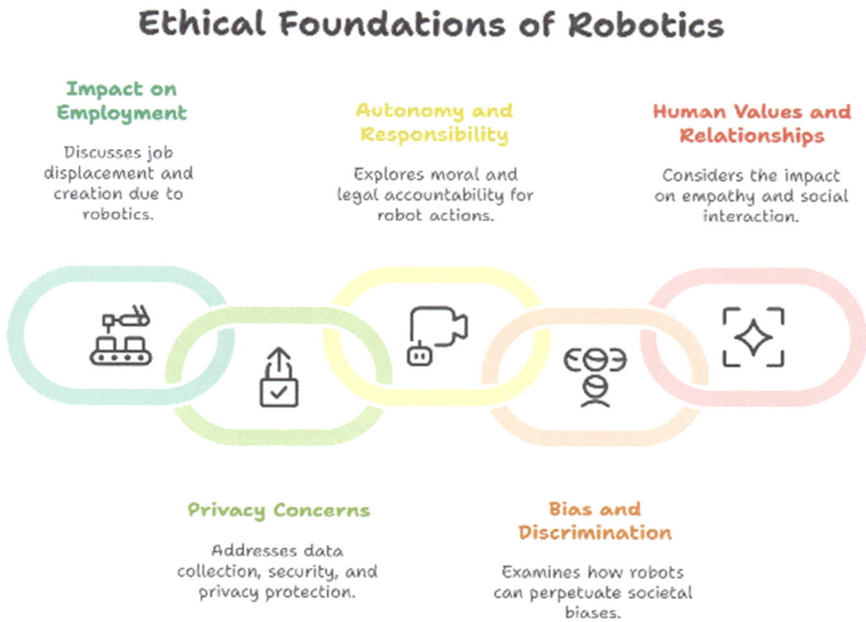

Fig. 12.4 Interconnected ethical foundations of robotics

robot deployment, Autonomy and Responsibility emphasises moral and legal robot actions, Human Values and Relationships considers their impact on empathy and social interaction, Privacy Concerns highlights data protection and robot surveillance, and Mitigation addresses unforeseen societal effects and safety measures, forming an interconnected framework for responsible robotics development.

Employers, government agencies, and the education sector must collaborate to implement skill development programs, which have been crucial in managing workforce change. These courses should cover both soft skills—which are unique to humans—and technical skills, such as how robot systems work.

12.5.2 Privacy, Safety, and Ethics

To maintain public trust, the use of robotic systems also presents serious privacy, safety, and ethical application issues that need to be considered.

Because it can gather a lot of information about human behaviour and actions, the large-scale sensing capability raises privacy concerns. To address these concerns, robust data protection measures, user consent systems, and technical controls such as local processing are necessary.

The aspect of safety involves both physical and functional safety, where the systems must be dependable, not to cause harm, but must be functional. Ethical theories are utilised in the design and implementation of solutions in order to uphold the relationship with human ideals.

12.5.3 *Accessibility and Digital Divide*

There is no evidence that the developers focused on architectural forms and principles when writing this specification. The ubiquity of robots presents both learning opportunities and challenges to access that must be addressed to ensure a fair distribution of technological advantage.

The implication of the digital divide is that economic access, geographical location, and demography can be constraining factors in accessing robotic technology. Educational curriculum adaptation must integrate robotics literacy while addressing diverse learning needs.

Accessibility design ensures robotic systems serve users with diverse abilities rather than creating new barriers. Cultural sensitivity recognises that systems must operate appropriately across diverse contexts with varying social norms.

12.6 Challenges and Future Directions

12.6.1 *Technological Challenges*

Contemporary robotics faces fundamental technological challenges that limit current capabilities and define research priorities for future development [19, 20].

Perception and Sensing remain critical bottlenecks for robotic autonomy, particularly in unstructured environments with variable conditions. Current sensor technologies struggle with integration across multiple modalities and real-time processing of high-dimensional sensory data.

Energy Efficiency constrains operational capability of mobile and autonomous systems, limiting mission duration and computational capability. Current battery technology provides insufficient energy density for many applications.

Reliable Autonomy requires systems that can operate safely across the full range of conditions encountered during deployment. Current systems often perform well in controlled environments but struggle with unexpected situations.

12.6.2 Emerging Technologies and Trends

Several emerging trends are reshaping robotics, offering new capabilities while creating novel technical and societal challenges [21].

Collaborative Robots represent a shift toward systems designed for productive, safe interaction with human workers. These systems require advances in safety technology and adaptive behaviour responding to human preferences.

Soft Robotics employs flexible materials and novel actuator designs to create systems that can safely interact with delicate objects while adapting to irregular environments.

Swarm Intelligence enables collections of simple robots to solve complex problems through distributed coordination and emergent behaviour, offering scalability and fault tolerance advantages.

Figure 12.5 illuminates the multifaceted advancements in robotics as of August 2025, with Artificial Intelligence and Machine Learning enabling robots to perform complex tasks and adapt to environments, Haptics and Tactile Sensing providing natural interaction capabilities, Computer Vision allowing robots to "see" and interpret surroundings, Advanced Materials and Manufacturing enhancing robot performance and durability, and Automation/Robotics showcasing diverse industrial applications. It also addresses challenges and future directions, highlighting hurdles and potential paths to shape the evolution of robotics technology.

12.6.3 Future Research Frontiers

Future robotics research is defined by ambitious challenges pushing current technology boundaries while addressing critical societal needs [22].

Extreme Environment Robotics addresses the need for systems operating reliably in conditions hostile to humans and conventional technology, including deep space, ocean depths, and extreme temperatures.

AI-enhanced decision making is referred to as the development of systems that can arrive at complex decisions in uncertain situations, ensuring safety and alignment with human values.

Based on the circular economy and the application of environmental considerations, Sustainable Robotics aims to address environmental impact by developing technologies that aid in the development of sustainability objectives.

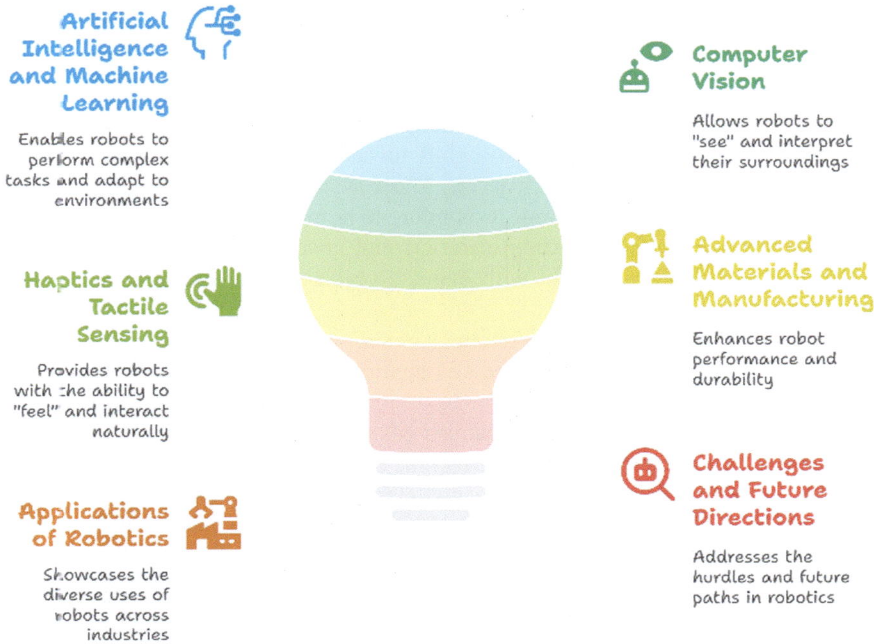

Fig. 12.5 Illuminating the future of robotics innovations

12.7 Conclusion

12.7.1 Key Insights and Contributions

Robotics and autonomous systems have significantly changed the autonomous production in industry, the nature of human–computer interactions, and the quality of scientific research. These systems are comprised of the integration of various existing technologies and techniques, including IoT-based sensors, artificial intelligence, and improved materials. The application market has a cognitive imagination that extends beyond the traditional manufacturing world, encompassing almost any industry of human life, and demonstrates incredible flexibility in overcoming various problems.

The addition of artificial intelligence has enabled the integration of robotic systems into unstructured learning systems, which are adaptable enough to act as special-purpose programming for automation, rather than being programmed with software written in traditional programming languages or modification languages. Nonetheless, the ramifications of society extend beyond technical aspects and include the transformation of the workforce, ethical decision-making, and equal access, which must be managed proactively.

12.7.2 Future Outlook and Implications

Autonomous systems to be realised next will still imply additional advances to your relations with technology and to our attitude to complicated problems. Higher autonomy grades will support a longer period of independent operation, whereas more opportunities to use human–robot collaboration will form flawless partnerships dynamically dividing the tasks according to their abilities and demands.

New domains are expected to emerge from the convergence of the latest technologies, including biotechnology, advanced materials, and quantum computing. It is essential to address human problems through the creation of sustainable technology, interdisciplinary research and development, and the careful evaluation of societal consequences.

As we move forward with autonomous solution-making using robotics, it is our duty to ensure that the progress we make respects human dignity, promotes equity, and fosters environmental sustainability, ultimately leading to a prosperous world for future generations.

References

1. Acharya DB, Kuppan K, Divya B. Agentic AI: autonomous intelligence for complex goals—A comprehensive survey. IEEE Access. 2025.
2. Bilal M, Ameer S, Awan AS, Mumtaz A, Gul S, Shahbaz N. Regulating the future: a holistic model for safe and ethical deployment of robotic systems in emerging industries. J Comput Biomed Inf. 2025; 9(01).
3. Spitale M, Axelsson M, Jeong S, Tuttösí P, Stamatis CA, Laban G, Lim A, Gunes H. Past, present, and future: a survey of the evolution of affective robotics for well-being. IEEE Trans Affect Comput. 2025.
4. Di Bello F, Rodriguez Peñaranda N, Marmiroli A, Longoni M, Falkenbach F, Le QC, Tian Z, Goyal JA, Colla Ruvolo C, Califano G, Creta M. In-hospital outcomes after robotic versus open radical nephroureterectomy. J Endourol. 2025;39(2):127–34.
5. Turja T, Kork AA, Ilomäki S, Hellstrand I, Koistinen AK. Care robot literacy: integrating AI ethics and technological literacy in contemporary healthcare. AI and Ethics. 2025;5(3):2623–40.
6. Murphy RR. Robots in science fiction unload assumptions about freight logistics. Sci Robot. 2025; 10(105):eaea7372.
7. Fosch-Villaronga E, Shaffique MR, Schwed-Shenker M, Mut-Piña A, van der Hof S, Custers B. Science for robot policy: advancing robotics policy through the EU science for policy approach. Technol Forecast Soc Chang. 2025;1(218): 124202.
8. Liang S, Rajora M, Liu X, Yue C, Zou P, Wang L. Intelligent manufacturing systems: a review. Int J Mech Eng Robot Res. 2018;7(3):324–30.
9. Irshad A, Farooq N, Kiran Q, Farooq A, Munir S, Kafeel S. Integration of robotics in healthcare management: a narrative review. Anaesth Pain Intensive Care. 2025;29(2):331–6.
10. Firoozi R, Tucker J, Tian S, Majumdar A, Sun J, Liu W, Zhu Y, Song S, Kapoor A, Hausman K, Ichter B. Foundation models in robotics: applications, challenges, and the future. Int J Robot Res. 2025;44(5):701–39.
11. Fan H, Liu X, Fuh JY, Lu WF, Li B. Embodied intelligence in manufacturing: leveraging large language models for autonomous industrial robotics. J Intell Manuf. 2025;36(2):1141–57.

12. Isaka S. Taxonomic robot identifiers: toward general classification and oversight for autonomous systems. IEEE Access. 2025.
13. Pescio M, Kundrat D, Dagnino G. Endovascular robotics: technical advances and future directions. Minim Invasive Ther Allied Technol. 2025;34(3):239–52.
14. Kaur N, Sharma A. Robotics and automation in manufacturing processes. In: Intelligent manufacturing. CRC Press;2025. p. 97–109.
15. Javaid M, Haleem A, Pratap Singh R, Rab S, Suman R, Kumar L. Utilization of robotics for healthcare: a scoping review. J Ind Integr Manag. 2025;10(01):43–65.
16. Moshayedi AJ, Khan AS, Geng K, Hu J, Kolahdooz A. Advancing agricultural practices: analyzing the role of robotics in corn farming. Int J Eng Trans A Basics. 2025;38:1517–32.
17. Velastegui R, Poler R, Diaz-Madronero M. Revolutionising industrial operations: the synergy of multiagent robotic systems and blockchain technology in operations planning and control. Expert Syst Appl. 2025;9: 126460.
18. Ferhataj A, Memaj F, Sahatcija R, Ora A, Koka E. Ethical concerns in AI development: analyzing students' perspectives on robotics and society. J Inf Commun Ethics Soc. 2025;23(2):165–87.
19. Wang G, Zhang C, Liu S, Zhao Y, Zhang Y, Wang L. Multi-robot collaborative manufacturing driven by digital twins: advancements, challenges, and future directions. J Manuf Syst. 2025;1(82):333–61.
20. Selvam DC, Devarajan Y, Raja T. Exploring the potential of artificial intelligence in nuclear waste management: applications, challenges, and future directions. Nucl Eng Des. 2025;1(431): 113719.
21. Urrea C, Kern J. Recent advances and challenges in industrial robotics: a systematic review of technological trends and emerging applications. Processes. 2025;13(3):832.
22. Krejčí J, Babiuch M, Suder J, Krys V, Bobovský Z. Internet of robotic things: current technologies, challenges, applications, and future research topics. Sensors. 2025;25(3):765.

Chapter 13
Edge Computing: Processing at the Edge

13.1 Introduction to Edge Computing

13.1.1 Motivation and Evolution from Cloud to Edge

To foster the growth on the clouds, users actively receive and obtain data, rather than passively receiving it, both at the point of selection and implementation. The architectural development between centralised cloud computing and distributed edge paradigms constitutes a key shift in the way in which information system computational resources are arranged, deployed and used in modern information systems. Such a move helps to overcome major constraints present in cloud-based architectures as they are implemented in new applications with latency requirements below 1 ms, greater privacy safeguarding, and autonomy of control [1].

Formal cloud computing systems have transformed the overall computing landscape at the enterprise level by offering virtually unlimited computer resources, high-tech software suite services, and economies of scale that have significantly reduced the operational costs of numerous applications. Nevertheless, the physical laws underlying data transmission have inherent limitations in cloud-based systems: the speed of light restricts the speed at which data can be transmitted between geographically separated locations, and unavoidable latency imposes inherent constraints that are incompatible with time-sensitive applications [2].

Demand for advertisements, such as autonomous vehicles, industrial automated systems, augmented reality interfaces, real-time medical monitoring, and other applications, requires response times in milli- or microsecond timescales, which cannot be reliably obtained when data must travel over networks to remote cloud data centres. Round-trip communication delays with servers in geographically remote locations in addition to processing delays and possible network congestion result in the aggregate response times larger than both the safety and performance specifications in these new application domains [3].

T. H. Sardar, *The Spectrum of Computer Science: Emerging Technologies and Trends*, Studies in Smart Technologies, https://doi.org/10.1007/978-3-032-19632-3_13

The astronomical growth of Internet of Things (IoT) devices has increased the vulnerability of cloud based architectures to scaling constraints [4]. It has been estimated that in the year 2030 there will be a total of more than 50 billion connected devices producing exabytes of data each day. Sending all of this data to centralised cloud systems would place burdensome network links on existing global capacity and invoke banishing operational cost. In addition, numerous applications of IoT rely on live local responses to sensor data, and cloud-based processing architectures are fundamentally not applicable to the real-time decision making process. There has arisen a need to focus privacy concerns, and a demand towards data sovereignty, by organisations and individuals wishing further control of sensitive information processing, claiming this edge computing adoption is contributing to counter-cyclical trends. The regulations, including the General Data Protection Regulation (GDPR) of the European Union, other national laws on data localization and industry-specific privacy regulations generate compliance requirements that encourage local data processing as opposed to transmission to distant remote cloud servers, potentially subjected to a different legal framework and cloud security standards.

This has made efficiency in the consumption of energy appear to have a higher profile following the growth that has been made in the area of environmental sustainability and costs incurred in utilization of energy in all areas. Transferring bulk data to remote cloud servers involves high energy usage on network infrastructure, compared to edge processing, as the former conducts large-scale analysis and transfers only processed products or pertinent summaries to decentralised systems.

13.1.2 Fundamental Principles of Edge Computing

Edge computing is based on various essential principles that contrast with the traditional centralised computing paradigm and define its features, advantages, and design specifics. These values help architects make informed choices about buildings and strategies to implement them in various fields of application [5].

The basic premise behind edge computing is Proximity-based Processing, which places computational resources physically close to both data sources and end users. This closeness decreases the latency of communication, achieves low network bandwidth usage, and facilitates real-time receptiveness to offer those applications that are sensitive to timeliness. Placement of edge resources optimally involves close consideration of the needs of use, constraints of network topology, and the availability of resources to minimise the benefits of performance against deployment and operation costs.

Distributed Intelligence doesn't have to focus on central processing a facility and allocate computational intelligence to numerous autonomous edge nodes. This monthly distribution increases the system's resistance to failures because it eliminates a single point of failure, and it enables the parallel processing of local data streams. To achieve effective distributed intelligence, advanced coordination elements are necessary, ensuring uniform system behaviour and adaptation to local circumstances and needs.

Data Locality places data processing needlessly close to the location of the data it operates on, eliminating the need to move data over long distances and ensuring quicker response times and reduced network bandwidth usage. This principle is especially vital when it comes to applications where high-volume data streams may emerge or where sensor inputs need to be processed as soon as possible. To achieve efficient data locality, proper algorithms are required to determine which data needs to be processed locally and which data needs to be sent to remote systems for centralised processing.

Edge systems ensure their autonomy and resilience, allowing them to continue operating in the event of a network crash or if the capacity to connect with centralised systems is capped or offline. Such independence is vital in the case of applications that are in a segmented area or are mobile or other highly sensitive infrastructure that require a constant presence of operation, notwithstanding network utilisation. To design autonomous edge systems, local resource capabilities, failure conditions, and graceful degradation approaches must be carefully considered.

Heterogeneity Accommodation acknowledges the fact that edge environments are comprised of a variety of device types, computational devices, communication protocols, and operational needs. The architectures of edge computing need to support this heterogeneity and offer homogeneous application programming interfaces and management opportunities. This can only be achieved with flexible middleware systems, standardised communication protocols, and adjustable resource management algorithms.

13.1.3 Comparative Analysis: Cloud, Fog, and Edge Computing Paradigms

Understanding the differences and similarities between edge computing, cloud computing, and fog computing is crucial for selecting the appropriate architecture for specific applications, as well as designing hybrid systems that leverage the strengths of all three methods [6, 7].

Cloud Computing offers centralised, scalable remote data centers that are available by means of internet connection. Cloud computing platforms are excellent to offer virtually limitless processing and storage environments, advanced software services, and economies of scale lowering the cost of large applications. Nonetheless, cloud computing fundamentally comes with network distance latencies, and needs a stable network connection, and might lack sufficient control over the location and processing of data to meet privacy-sensitive applications.

The cloud model is optimisation towards applications that can survive latencies in moderate system requirements, vast system demands that the local system is incapable of reaching, data aggregation and analysis to a central service, and access to specialised software or machine learning environments. Typical sample applications are big data analytics, time-dressed data storage, detailed machine learning training, and services to global populations of users.

Fog Computing is a mid-level architecture to deploy cloud computing to users who are even nearer to the device, in what is a distributed data chain of computing nodes in the form of a hierarchy. The use of numerous processing layers, often between local edge devices and regional fog nodes, up to centralised resources in the clouds is a common feature of fog architectures. The hierarchical structure allows applications to trade off latency versus computational resource use through delivering time-sensitive traffic at fog nodes and making use of cloud resources at cloud nodes to perform complex analytics and provide long-term storage capabilities. Figure 13.1 provides a computing comparison of cloud, fog, and edge computing.

Edge Computing brings computational resources to the network edge, where data is gathered and consumed, close to the network edge and the data sources and the end users. It minimises latencies, bandwidth needs and because it can experiment autonomously when the network suffers issues. Edge computing is best suited for

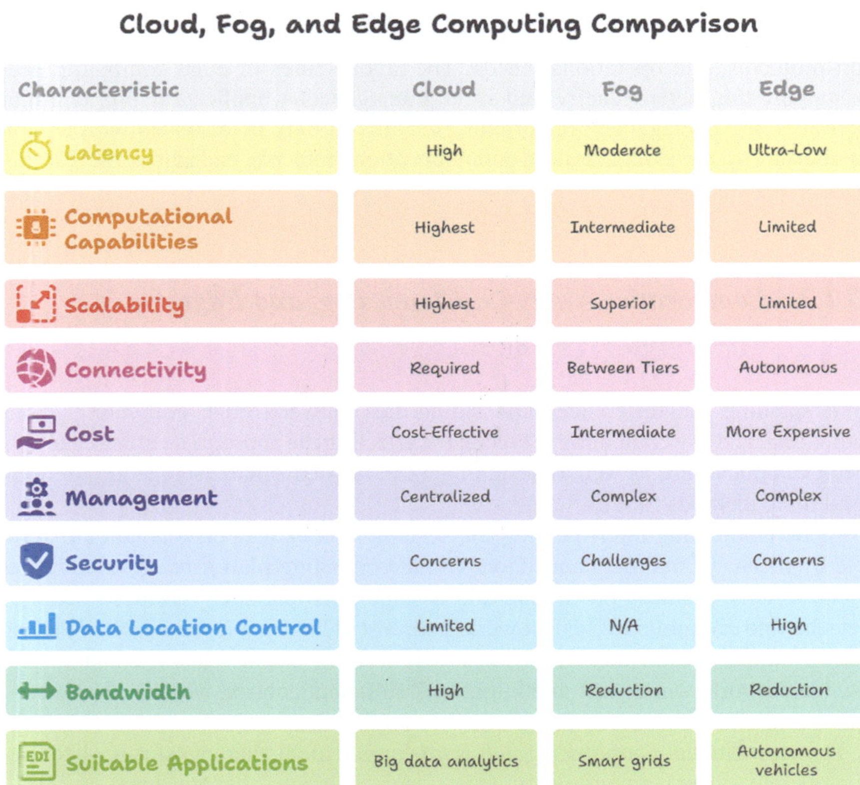

Cloud, Fog, and Edge Computing Comparison

Characteristic	Cloud	Fog	Edge
Latency	High	Moderate	Ultra-Low
Computational Capabilities	Highest	Intermediate	Limited
Scalability	Highest	Superior	Limited
Connectivity	Required	Between Tiers	Autonomous
Cost	Cost-Effective	Intermediate	More Expensive
Management	Centralized	Complex	Complex
Security	Concerns	Challenges	Concerns
Data Location Control	Limited	N/A	High
Bandwidth	High	Reduction	Reduction
Suitable Applications	Big data analytics	Smart grids	Autonomous vehicles

Fig. 13.1 Comparative analysis of cloud, fog, and edge computing paradigms

scenarios that demand extremely short response times, can run in conditions with limited bandwidth, and process sensitive data that is not intended to leave the local facility, or in conditions with unguaranteed network access.

These paradigm trade-offs include balancing the latency requirements, the computational capabilities, scalability requirements, connectivity requirements, and cost considerations. Bare in the sky architectures have unlimited computing power and application complexity at the cost of larger and more imprecise latencies and connectivity requirements. Pure edge designs are the least latent, highest autonomy, but reduce latency, and can create more complex deployments. Fog computing offers trade-offs that fall somewhere in the middle, but it requires coordination between several architectural layers.

13.2 Edge Devices and Architectures

13.2.1 Characteristics and Taxonomy of Edge Devices

The basic computational units of a distributed edge graph, namely edge devices, are endowed with specific features that set them apart in combination with traditional computing platforms, as well as their capabilities and limitations in distributed systems. Such devices should be able to strike a good compromise between the computational needs and the firm limitations imposed by operational conditions, energy availability, and form factor considerations [8].

Processing Power and Energy Optimisation

Modern edge devices utilise more sophisticated System-on-Chip (SoC) architectures, which combine one or more general-purpose CPU cores with dedicated signal processing (SPU) units, artificial intelligence (AI) accelerators, and even dedicated input/output controllers, all integrated into a single chip. They consume less power because these integrated designs have optimised data paths, shared memory structures, and dynamic power management functions to vary resource utilisation to current computational patterns.

Dedicated AI accelerators, such as Tensor Processing Units (TPUs), Neural Processing Units (NPUs), and Field-Programmable Gate Arrays (FPGAs), possess high-performance machine learning capabilities coupled with energy efficiency that is significantly higher than that of standard general-purpose processors. These accelerators enable edge devices to perform complex artificial intelligence functions, such as computer vision, natural language processing, and predictive analytics, on-site without requiring external network connectivity, thereby operating in the cloud.

Some of the most modern power management methodologies include Dynamic Voltage and Frequency Scaling (DVFS), which variably adjusts the processor's operating point depending on the computational load; intelligent sleep and hibernation, which significantly minimise power during idle conditions; and intelligent task scheduling, which optimises energy usage in the presence of multiple processing elements. Energy collection can leverage power sources in the environment, such as solar radiation, thermal energy, vibration, and radio frequency emissions, due to the possibilities of energy harvesting.

Connectivity and Communication Architectures

The edge device shall be able to interface successfully in a heterogeneous communication space comprising a variety of wireless and wired protocols, different Quality of Service (QoS) features, and patchy connectivity. Such diversity requires flexibility in communication architecture, which must ensure the operation of network resources within the context of available resources and preserve the application's functionality.

There are often various categories of wireless communication technologies that include short-range communications like Bluetooth Low Energy (BLE), Zigbee, and WiFi used in a local area, cellular communications including Long-Term Evolution (LTE) and 5G in a wide area, and Low-Power Wide-Area Networks (LPWAN) like LoRaWAN and Narrowband IoT (NB-IoT) long-range communications.

Protocol interoperability is a paramount necessity for edge devices that must be able to interact with various systems, utilising different data formats, communication protocols, and security systems. Normative constructive schemes, such as Message Queuing Telemetry Transport (MQTT), Constrained Application Protocol (CoAP), and HTTP-based Representational State Transfer (REST) APIs, share common communication frameworks that enable interoperability between various types of devices and vendor implementations.

13.2.2 Device Categories and Deployment Models

The variety of applications for edge computing has led to innovations in various device categories, all tailored to specific operational and computational needs, and implementation environments [9].

Microcontrollers and Embedded Systems

Microcontrollers are the most resource-constrained type of edge device, with restricted processing power, memory bandwidth, and energy allocation, yet they offer the fundamental computational resources needed to interface with sensors, perform simple data processing, and communicate. These are designed to be optimised to run on very low power masses, and can sometimes last for years of battery-powered operation, but require periodic sensing and communication.

Modern microcontrollers also feature more sophisticated features, such as hardware cryptographic accelerators to aid in security functions, high-resolution analogue-to-digital converters to interface with sensors, and communications peripheral support for multiple protocols simultaneously, as well as other low-power design features that enable extended autonomous operation. Modern microcontrollers have

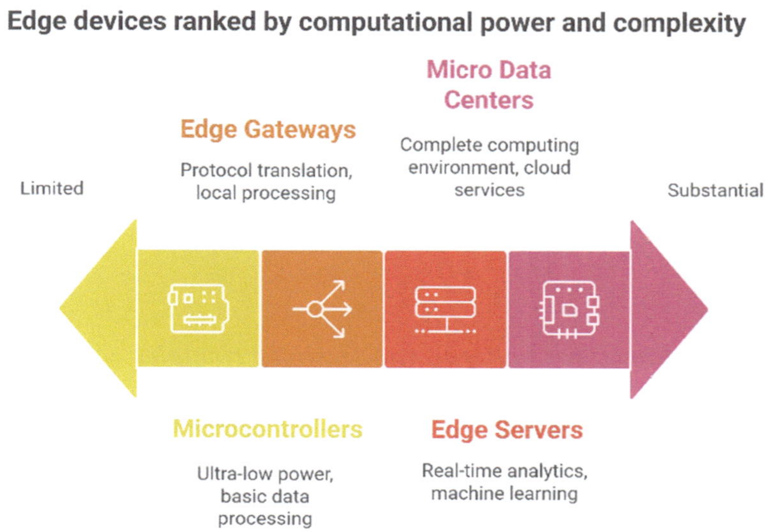

Fig. 13.2 Device categories and deployment models of edge computing

the capability of AI acceleration that can be used to perform simple machine learning inference activities like keyword recognition, sensor anomaly detectors, and pattern classifications. Figure 13.2 Visualises the device Categories and Deployment Models of Edge Computing.

Edge Gateways and Industrial Computing Platforms

Edge gateways are intermediate devices that act as a union of information readings across sensors or edge devices, perform local processing, and serve as a gateway to broader networks or cloud solutions. The important distinction is that these devices will often have significantly more power than microcontrollers, yet require industrial-grade reliability and environmental resilience.

Protocol translation functions enable edge gateways to facilitate translation between disparate communication technologies, seamlessly transitioning between older industrial communications protocols, such as Modbus or Controller Area Network, and more modern communications, like IP-based networks. This possibility of translation is important in terms of incorporating old industrial processors into new edge computing architectures, rather than replacing the costly equipment.

Edge Servers and Micro Data Centres

The edge servers have high computational power, similar to conventional data centre devices, yet are geared towards implementation in remote locations with limited infrastructure facilities. These systems facilitate advanced applications, including real-time video analytics, advanced machine learning inference and local cloud services, but with reliability and performance rivalling those of traditional data centre applications.

Micro data centres are entire computing scenarios placed in representative locations at the edge, comprising a conglomeration of computers, storage hardware, networking, power conditioning, and environmental control in a self-contained enclosure within a compact compartment. These systems enable them to deploy cloud-like services at the edge without compromising observability or the ability to manage their systems in a manner reminiscent of traditional data centres.

13.2.3 Architectural Patterns and Design Principles

Architecture for Edge computing designs includes various organisational designs that are optimised according to application capture, operational, and performance goals [10].

Hierarchical Architectures

Hierarchical edge computing architectures can be likened to the organisation of computational resources into numerous layers that have various capabilities and responsibilities, moving from resource-constrained edge devices all the way up to centralised cloud resources. With this organisation, applications are allowed to optimise where to place their operations as per the latency requirements, computer needs, and characteristics of data.

Common types of hierarchical implementations are device-level processing with real-time response processing and simple filtering, local gateway or fog node processing with aggregation, basic processing, intermediate analysis of data, regional edge data center processing with more complex processing and temporary storage, and shared cloud resources and long-term storage, advanced analytics, and global coordination functions.

Peer-to-Peer Distributed Architecture

Distributed architectures make hierarchical architectures more uniform by spreading computational capabilities among the edge nodes, allowing peer-to-peer computation, and removing dependencies and possible bottlenecks found in hierarchical organisations. This model grants the system more operational freedom and strength, although it demands more advanced coordination and may make resource administration more challenging.

Distributed consensus algorithms can coordinate peer edge nodes without the need for centralised control mechanisms, whereas distributed storage systems offer redundancy in the availability and consistency of data to multiple nodes. Such systems should tackle complex problems, such as network partitions, node failures, and address transaction operation congestion.

13.3 Applications and Use Cases

New areas of application demonstrate the game-changing aspect of edge computing in multiple industries, where real-time intelligence, reliability, and privacy are valued. The following subsections examine major verticals where edge computing technologies are rapidly transforming operational models and delivering next-generation capabilities [11, 12].

13.3.1 Internet of Things and Cyber-Physical Systems

Internet of Things can be considered amongst the indicators of edge computing diffusion which demonstrate the greatest growth, with billions of gadgets connected and producing vast amounts of data that needs smart computation and prompt examination. Edge computing will uniquely allow IoT applications to be real-time responsive, significantly conserve bandwidth, and continue functioning when a network crashes without violating privacy and cuts operation expenses.

Smart Infrastructure and Urban Systems

Applications of smart cities use edge computing to support responsiveness in traffic management, environmental tracking, the citizen security system, and municipal services of urban infrastructures. Distributed edge computing allows any city to take real-time data of thousands of sensors and cameras and coordinate responses in multiple departments and services areas.

Traffic management systems deploy edge computing capabilities at intersection controllers, traffic cameras, and roadside sensor units to optimise traffic flow, detect incidents, and coordinate emergency responses. Local processing enables immediate signal adjustments based on current traffic conditions while maintaining coordination with city-wide traffic optimisation systems and regional transportation planning authorities.

Environmental monitoring networks utilise edge computing at air quality sensors, noise monitoring stations, and meteorological equipment to provide real-time environmental data while identifying pollution sources and health hazards. Local processing capabilities enable immediate alerts for dangerous conditions while contributing data to city-wide environmental management and public health protection systems.

Industrial IoT and Manufacturing Systems

Industrial IoT applications represent some of the most demanding edge computing use cases, requiring real-time processing of sensor data for equipment monitoring, process control, and safety systems. Edge computing enables industrial applications to achieve millisecond response times required for process control while maintaining operational capability during network connectivity disruptions.

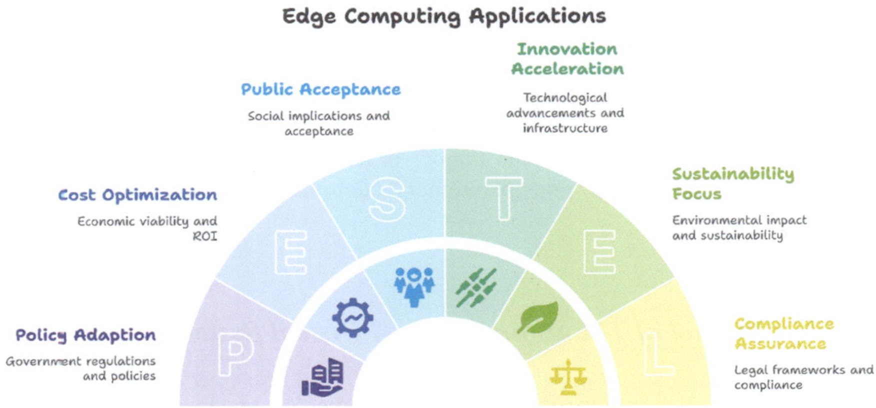

Fig. 13.3 The applications of edge computing

Figure 13.3 provides the applications of edge computing.

Predictive maintenance systems utilise edge computing to continuously monitor equipment health through vibration sensors, thermal imaging systems, acoustic monitoring equipment, and other sensor modalities. Local machine learning models analyse sensor data streams in real-time to detect early indicators of equipment degradation, enabling proactive maintenance scheduling before failures occur.

Process control applications need the definite processing that is able to ensure the maximum response time of process to safety-critical or quality-critical processes. Edge computing allows the control systems to react to the changes in the processes under the prescribed time constraints and to interact with the systems of optimisation of the plant that work on a scale longer time horizons.

13.3.2 Autonomous Systems and Robotics

Edge computing has emerged as a core component to the recent robotics applications and other autonomous systems in need of real-time decision support, adaptive behaviours, and safe operation in unpredictable in dynamic environments. Combination of edge processing and robot systems allows them to act autonomously and offers the computational power required to perform difficult processing on sensors and artificial intelligence algorithms.

Real-Time Decision-Making and Control

The types of robotic use require predictable response times, which are not feasible with cloud-processing because the network will be slow and the connections will be erratic. Edge computing allows robots to understand sensor data on the spot and makes critical decisions, but still retain connection with remote systems to perform functions that require higher-level planning and optimisation.

Edge computing takes sensor input (cameras, Light Detection and Ranging (LiDAR), radar, and inertial measurement unit) and real-time computations are used to develop environmental maps, identify hazards, and maneuver safe paths via full autonomy in navigation systems. Local processing will allow responding to dynamic barriers and conditions of the environment in time and retain good, efficient mobility properties.

AI accelerators used in computer vision applications in robotics accelerators are used to incur minimal latency recognition and pose estimation using edge data as well as understanding capabilities of the scene. The above capabilities will allow the robots to manipulate objects, to interact with humans in a safe way, and to navigate in dynamic environs without the need to transmit to the remote processing systems with high-bandwidth video images.

Human–Robot Collaboration

The specifics of collaborative robotics work demand a high level of knowledge of human wishes and needs, the constant monitoring of safety, and adaptability of behaviour, which reacts to human preferences and work styles. Edge computing, from this viewpoint, enables such applications to perform natural, reactive interactions while sustaining safety parameters and operational performance.

The safety monitoring systems can utilise edge computing to continuously observe the interactions between robots and human beings. By identifying possible collision risks, the robot's behaviour can automatically adjust to maintain a safe distance and reduce its movement speed. These systems should be able to operate with very low latency, thereby avoiding accidents, while still allowing for effective working efforts.

13.3.3 Healthcare and Biomedical Applications

Healthcare apps are among the most privacy-aware and the safety-critical uses with a strong requirement to monitor in real-time, be able to react to critical conditions and ensure data protection whilst exposing applications to a high degree of reliability critical to patient safety.

Remote Patient Monitoring

Remote patient monitoring services make use of edge computing to analyse vital signs, activity patterns and other health indicators in real-time and issue instant warnings about life-threatening health conditions. The local processing would allow real time measurement of health and reduce relay of sensitive health information to the outside world.

Wearable health sensors are based on the edge AI functionality to identify irregular heart rates and events of falls, seizures, and other medical crises and send an immediate notification to patients and healthcare professionals. Local processing can also provide fast emergency response and safeguard privacy related to patients by transmitting certain data selectively and performing an analysis on the device itself.

Privacy-Preserving Health Data Processing

Healthcare data include some of the most sensitive personal data such as medical history, diagnostic outcomes, genetic data, behavior assessment, and biometric data. This information is required to be safeguarded by elaborate regulatory regimes such as The Health Insurance Portability and Accountability Act (HIPAA) in the United States and other privacy laws in other parts of the world.

To meet healthcare privacy needs, edge computing will be able to process sensitive data locally at healthcare facilities or on personal devices, minimising exposure to data transmission and offering healthcare organisations more control over the use of protected health data. It should be noted, however, that its implementation will require careful consideration of data flows, access controls, and audit capabilities.

13.4 Security, Privacy, and Trust Considerations

The decentralisation of edge computing brings both challenging security, privacy, and trust needs that cannot be easily addressed in the traditional centralised, cloud-based architectures. Because gravity and location bring computation to data sources and end users, many thousands, or even millions, of edge devices with varying security requirements, different network environments, and distinct management paradigms now comprise the attack surface [13–15].

13.4.1 Distributed Security Challenges

The threat landscape in edge computing is a complex, distributed threat surface composed of constrained resource devices and varied connectivity topologies. Knowledge about these security issues can help build a robust edge computing architecture that can safely be used in seemingly hostile environments.

Device Security and Hardware Vulnerabilities

The edge devices are the most susceptible units in a distributed computing structure due to resource limitations, the multiplicity of manufacturing requirements, and implementation complications. The issues of device security include the vulnerability of hardware, integrity of firmware and lifecycle management of heterogeneous populations of devices.

The hardware threat to edge devices is often a result of cost pressure that seeks to reduce the cost to favour functionality over more extensive security features. Most edge devices do not have Hardware Security Modules (HSMs), secure boots or tamper-resistant enclosures allowing them to be physically attacked. Cryptographic keys or sensitive information of poorly secured devices can potentially be extracted by side-channel attacks, power analysis, and electromagnetic analysis.

Many edge device vendors and worldwide supply chains pose a risk to supply chain security, with bad actors having the potential to place hardware backdoors or firms with web-based attacks injecting firmware or installing malware in the manufacturing or distribution cycle. Modern semiconductor supply chains are too complicated to engage in complete protection verification, which is quite costly.

Network Security and Communication Protection

Edge computing systems are highly dependent on network connectivity to coordinate, synchronise data and control it remotely, posing a variety of attack patterns that involve network security and integrity in data transmission. The network technology diversity presents different security models and different threat profiles, which should be handled interactively.

Man-in-the-middle attacks pose a significant threat to edge computing devices, which can be connected via untrusted networks or through third-party-controlled intermediate devices. They might introduce unauthorised code that can destroy information or disrupt the functionality of the cloud services, allowing network attackers to compromise the edge devices and the cloud services.

13.4.2 Privacy Preservation Mechanisms

The inherent promise of improving privacy through local data processing, as offered by edge computing activities, will need to be accomplished via advanced technical deployments in response to the intricate data streams and processing demands of distributed networks.

Data Locality and Minimisation

The compliance needs of edge computing systems, driven by geographic data residency requirements, including regulations such as the GDPR, laws on data localisation, and sector-specific privacy requirements, present multifaceted compliance burdens that may involve processing and storing data with disparate jurisdictional requirements. Organisations need to be aware of and comply with different national and regional data handling, data storage, and transfer regulations between countries and regions.

Geographic data residency requirements. Edge computing can support sensitive data processing directly at the required geographic level, provided that the architecture is designed to guarantee that data processing, intermediate data, and derived insights are all within the proper jurisdiction. The situation is complicated as edge computing systems that can cut across countries or cloud-based services being offered across jurisdictions.

Advanced Cryptographic Techniques

Homomorphic encryption permits mathematical calculations on encrypted data without decryption such that it is possible to compute a privacy-preserving computation in a distributed system. Although modern homomorphic encryption protocols have an excessive computational cost, algorithm and hardware acceleration are making them more usable in edge computing devices.

Secure Multi-Party computation (SMPC) permits several parties to compute functions together and in a way that does not expose their inputs to other parties. This suits edge computing environments when more than two organisations are required to work together to conduct data analysis without infringing individual privacy rights.

13.4.3 Trust and Authentication Frameworks

Trust relationships in the distributed edge computing systems need advanced authentication and authorisation systems, which can survive with occasional connectivity and fluctuating resource availability.

Decentralised Identity Management

The identity management systems that use blockchain or distributed trust can allow edge computing systems distributed trust and verification capabilities and do not require centralised authorities and constant connectivity to cloud services. These systems have the capability to verify the identity of devices, check firmware, and maintain audit trails throughout distributed edge infrastructure.

Self-sovereign identity solutions enable individuals and organisations to be able to manage their identity details, and at the same time be able to verify and authenticate edge computing services. This not only solves privacy issues involving monitoring by identity providers it also offers needed security features in distributed systems.

Zero-Trust Security Models

Zero-trust security architectures also assume that the network location, devices, and users should not be granted any trustworthy access, and all access requests are to be verified and approved, which is a continuous process. Along the lines of zero-trust in edge computing aiming to enforce profile-verifying, attestation, and constant monitoring over distributed infrastructure is needed.

Micro-segmentation Methods establish finer security control boundaries within edge computing systems, which contain limited opportunities of security breaches and accords defence-in-depth defence. The technologies of Software-Defined Networking (SDN) and Network Function Virtualization (NFV) make it possible to create and manage the security boundaries dynamically.

13.5 Emerging Trends and Future Directions

13.5.1 Artificial Intelligence at the Edge

Artificial intelligence and edge computing convergence is one of the most important advances in distributed computing, as it allows making intelligent decisions on the network edge and meets latency, bandwidth, and privacy needs that cannot be fulfilled by centralised AI solutions.

Model Optimisation and Deployment

Model compression systems allow highly complex AI models to run on resource-limited edge devices by scaling back model size, computation needs, and memory consumption at the expense of a tolerable amount of accuracy and functionality. Quantisation is used to cut the accuracy of the neural network weights and activations, pruning is used to eliminate unnecessary connections and neurons, and knowledge distillation is used to transfer knowledge of intricate models to more practical representations.

Neural Architecture Search (NAS) is an automated hardware-efficient and performance-driven design pipeline that is used to design efficient neural network architectures. NAS has a chance to find new architectures providing a trade-off balance between accuracy and choice of computational efficiency, which might be better than manually designed networks in edge computing applications.

Federated Learning Paradigms

Federated learning allows collaborative machine learning between distributed edge devices without the concentration of sensitive training information. The model is also trained only on local data and then only model updates are sent to sharing with central aggregation servers, and this is a highly privacy-conscious approach that allows more accurate and representative models to be developed by allowing access to diverse and distributed datasets.

On-device training capabilities enable edge devices to continuously adapt and improve their AI models based on local data and changing conditions. This proves particularly valuable for personalisation, adaptation to local environmental conditions, and situations requiring continuous model updates without cloud connectivity.

13.5.2 5G and Beyond: Network Integration

The development of wireless communications technologies, specifically 5G and the upcoming 6G, introduces new opportunities and requirements for edge computing architectures, while also laying the groundwork for a network infrastructure that may support distributed edge computing on a global scale [16].

Ultra-Reliable Low-Latency Communications

Latency on 5G networks is significantly lower than in other earlier-generation wireless networks, and in theory, the latency of Ultra-Reliable Low-Latency Communication (URLLC) services can be reduced to 1 ms. This functionality enables new categories of real-time applications that require a real-time response to varying conditions.

Network slicing technology enables the creation of multiple virtual networks with distinct attributes on a shared physical infrastructure, allowing network operators to deliver customised connectivity services tailored to meet specific edge computing applications and performance requirements.

Multi-access Edge Computing Standards

Standardisation of Multi-access Edge Computing (MEC) through the work of standardisation bodies, such as the European Telecommunications Standards Institute (ETSI), is defining standardised frameworks for implementing edge computing capabilities within telecommunications network infrastructure. These versions enable interoperability among various network operators and equipment providers, providing a consistent interface for application programmers.

13.5.3 Sustainability and Green Edge Computing

The innovation of sustainable edge computing architectures is being inspired by environmental concerns, the immediate environmental impact of edge computing infrastructure, and how it can be used to support more sustainable practices in various fields.

Energy-Efficient Architectures

Energy-conscious edge architectures should be able to strike a balance between computational ability and environmental sustainability through efficient design and intelligent power control, as well as by incorporating renewable energy sources. These involve the design of ultra-low-power processors, energy-efficient communication systems, and intelligent energy scheduling methods.

Carbon-neutral edge computing projects include integrating renewable energy, heat recovery, and the principles of a circular economy into the lifecycle of edge computing equipment. These solutions not only address the environmental impacts of edge computing but also demonstrate ways to achieve sustainable digital infrastructure.

As illustrated in Fig. 13.4, edge computing is strengthened by multiple complementary dimensions.

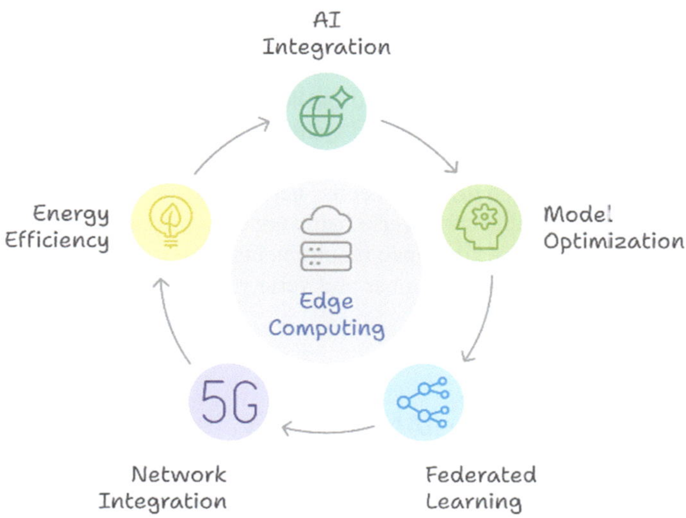

Fig. 13.4 Key dimensions of edge computing, showcasing its integration with AI, model optimisation, federated learning, 5G connectivity, and energy efficiency

13.6 Research Challenges and Future Directions

With edge computing developing, its usefulness is increasingly influenced by integration with new technologies that enhance its intelligence, connectivity, and sustainability. There are three key dimensions that the future of edge ecosystems takes, as illustrated in the subsequent sections, namely AI at the edge, advanced networking, and green computing [17–20].

13.6.1 Standardisation and Interoperability

The edge computing ecosystem is still highly fragmented as many vendors, platforms, and proprietary standards are competing to gain adoption in the market. Such fragmentation poses significant problems to organisations wishing to implement edge computing, and developers trying to develop applications that operate in a wide variety of edge environments.

Standardisation of protocols should govern the multiple communication protocols, data formats, and program interfaces implemented on various edge computing platforms. The common standards are being developed by standard organisations such as IEEE, ETSI and the Linux Foundation, though widely adopted by the industry is still yet to be achieved due to competition and technical complexity.

13.6.2 Resource Management and Orchestration

The optimal use of resources in distributed and heterogeneous edge computing infrastructures is a multifaceted algorithmic and system challenge that requires innovations in scheduling algorithms, modelling, and resource usage practices.

To make dynamic scheduling decisions, heterogeneous resource optimisation should consider the available and diverse hardware capabilities, network conditions, energy requirements, and performance requirements. This complexity is increasing exponentially with the growing number and variety of edge resources, which have varying degrees of deployment.

13.6.3 Ethical and Societal Implications

Fairness, accountability, transparency, and social impact are significant considerations associated with edge computing deployments, which must be carefully considered and examined in future research.

Edge AI systems should ensure their algorithmic fairness to mitigate all potential biases in training data, model design, and deployment situations that could lead to discriminatory outcomes. This is especially challenging when distributed data is being trained with possibly varying patterns of bias in its distribution across geographic locations and demographic subgroups.

Digital equity issues should concern the ways in which edge computing implementations can exacerbate or mitigate existing disparities in access to technology and digital services. To ensure fair access to the benefits of edge computing, it is necessary to plan and, if necessary, implement policy intervention measures to prevent the creation of new digital divides.

13.7 Conclusion

Edge computing is an essential architectural change in distributed computing systems, where computation is pushed to the edge in proximity to data sources and end-users, to satisfy latency, bandwidth, privacy, and reliability demands that cannot be reasonably satisfied using solely a centralised model of cloud computing. This change makes possible new classes of applications and services as it generates new challenges and opportunities at technological, organisational, and societal levels.

The security and privacy context of edge computing encompasses multifaceted, dynamic needs that necessitate multifaceted solutions to the various aspects of distributed, resource-constrained, and potentially adversarial deployment environments. Existing security models that serve centralised systems will have to be modified and improved to meet the increased attacker surface, wider and more varied trust relationships and the dynamic nature of the operations imposed by edge computing deployments.

The intersection of edge computing and artificial intelligence, novel wireless networking technologies, and new paradigms like quantum computing provides new opportunities in distributed intelligent systems. Such convergent technologies permit applications such as autonomous vehicles, smart cities, industrial automation, and personalised healthcare, which could not be done with traditional computing architectures.

Edge computing deployments are shifting towards a greater focus on sustainability considerations, prompting the creation of more energy-efficient hardware, the integration of renewable energy, and carbon-neutral operations. The possibility of edge computing to empower more sustainable approaches in other areas, such as smart buildings, transportation optimisation, and industrial efficiency, provides further incentive to pursue edge computing with environmental concerns in mind.

The future of the research path will determine the direction in which edge computing will continue to evolve on multiple levels. Achieving the scale and complexity of future edge computing deployments will require autonomous edge systems that are self-configured, self-optimised, and self-healing with minimal human intervention. This necessitates further developments in artificial intelligence, automated reasoning, and distributed system management.

The future success of edge computing will not only rely on technological developments, but also on broader concepts such as standardisation, governance, privacy protection, security assurance, and social impact. This also entails resolute cooperation between technologists, policymakers, industry actors, and communities at risk, so that edge computing emerges in a manner that is more beneficial to society without compromising individual rights and values.

As edge computing continues to evolve and expand worldwide, it will become a more central part of the digital infrastructure, enabling new uses and services that scale human capabilities and quality of life. The problems and opportunities discussed in this chapter will continue to motivate computing innovations and research in this paramount field of computer science, which will ultimately shape the future of distributed computing and its transformative impact on society.

References

1. Angel NA, Ravindran D, Vincent PD, Srinivasan K, Hu YC. Recent advances in evolving computing paradigms: cloud, edge, and fog technologies. Sensors. 2021;22(1):196.
2. Nezami Z, Chaniotakis E, Pournaras E. When computing follows vehicles: decentralized mobility-aware resource allocation for edge-to-cloud continuum. IEEE Internet of Things Journal. 2025.
3. Sharma M, Tomar A, Hazra A. Edge computing for industry 5.0: fundamental, applications, and research challenges. IEEE Internet Things J. 2024;11(11):19070–19093.
4. Mallik S, Qin H, Nandi S, Alam MY, Roy A, Sardar TH, editors. Applications of computational learning and IoT in smart road transportation system. Springer Nature;2025.

5. Sharma M, Tomar A, Hazra A (2024) Edge computing for industry 5.0: fundamental, applications, and research challenges. IEEE Internet Things J. 2024; 11(11):19070–19093. https://doi.org/10.1109/JIOT.2024.3359297.
6. Huaranga-Junco E, González-Gerpe S, Castillo-Cara M, Cimmino A, García-Castro R (2024) From cloud and fog computing to federated-fog computing: a comparative analysis of computational resources in real-time IoT applications based on semantic interoperability. Future Gener Comput Syst. 2024 159:134–150. ISSN 0167-739X. https://doi.org/10.1016/j.future.2024.05.001.
7. Balabantaray SK, Nayak MR, Panda B, Gourisaria MK, Basu A, Sardar TH. An improved image watermarking technique in spatial domain using histogram equalization and compass edge detection with adaptive alpha blending technique. In: 2025 4th OPJU international technology conference (OTCON) on smart computing for innovation and advancement in industry 5.0. IEEE;2025. p. 1–8.
8. Uddin R, Kumar SA, Chamola V. Denial of service attacks in edge computing layers: taxonomy, vulnerabilities, threats and solutions. Ad Hoc Netw. 2024;152:103322. ISSN 1570-8705. https://doi.org/10.1016/j.adhoc.2023.103322.
9. He C, Meng Q, Xu X, Chen P, Liang P, Cao J. An improved lightweight residual network model deployed on the edge device for the unsupervised cross-domain fault diagnosis. Expert Syst Appl. 2026; 296(Part B):129106. ISSN 0957-4174. https://doi.org/10.1016/j.eswa.2025.129106.
10. Pillai SE, Paramasivam S, Polimetla K, Dhanasekaran S, Agrawal KK, Yadav SP, Logeshwaran J, Gatto G. Optimizing resource management using hybrid metaheuristic algorithm for fog layer design in edge computing. Syst Soft Comput. 2005; 7, 200323. ISSN 2772-9419. https://doi.org/10.1016/j.sasc.2025.200323.
11. Liang S, Jin S, Chen Y. A review of edge computing technology and its applications in power systems. Energies. 2024; 17(13):3230. https://doi.org/10.3390/en17133230.
12. Zengshan YIN, Changhao WU, Chongbin GUO, Yuanchun LI, Mengwei XU, Weiwei GAO, Chuanxiu CHI. A comprehensive survey of orbital edge computing: systems, applications, and algorithms. Chin J Aeronaut. 2025; 38(7):103316. ISSN 1000-9361, https://doi.org/10.1016/j.cja.2024.11.026.
13. Shafee A, Hasan SR, Awaad TA. Privacy and security vulnerabilities in edge intelligence: an analysis and countermeasures. Comput Electr Eng. 2025; 123(Part B):110146. ISSN 0045-7906, https://doi.org/10.1016/j.compeleceng.2025.110146.
14. Latif R, Ahmed MU, Tahir S, et al. A novel trust management model for edge computing. Complex Intell Syst. 2022;8:3747–63. https://doi.org/10.1007/s40747-021-00518-3.
15. Niu S, Kong W, Chen L, Zhou X, Wang N. Privacy-preserving and verifiable federated learning with weighted average aggregation in edge computing. J Netw Comput Appl. 2025; 240:104201, ISSN 1084-8045. https://doi.org/10.1016/j.jnca.2025.104201.
16. Carvalho G, Cabral B, Pereira V, et al. Edge computing: current trends, research challenges and future directions. Computing. 2021;103:993–1023. https://doi.org/10.1007/s00607-020-00896-5.
17. Arroba P, Buyya R, Cárdenas R, Risco-Martín JL, Moya JM. Sustainable edge computing: challenges and future directions. J Softw: Pract Exp. 2024;54:2272–96. https://doi.org/10.1002/spe.3340.
18. Zhang Y, Feng J. Towards a smart and sustainable future with edge computing-powered internet of things: fundamentals, applications, challenges, and future research directions. J Inst Eng India Ser B. 2025;106:785–804. https://doi.org/10.1007/s40031-024-01186-w.
19. Andriulo FC, Fiore M, Mongiello M, Traversa E, Zizzo V. Edge computing and cloud computing for internet of things: a review. Informatics. 2024; 11(4):71. https://doi.org/10.3390/informatics11040071.
20. Loutfi SI, Shayea I, Tureli U, El-Saleh AA, Tashan W. An overview of mobility awareness with mobile edge computing over 6G network: challenges and future research directions. Results Eng. 2024; 23:102601. ISSN 2590-1230. https://doi.org/10.1016/j.rineng.2024.102601.

Chapter 14
The Metaverse Revolution: Shaping Workspaces, Learning, and Digital Interaction

14.1 Introduction

The metaverse is an existing virtual mechanism that is reshaping how we work, learn, and socialise. It is a transitional move to what would be referred to as cyber–physical–social convergence, as it is defined as a recurring, shared, and highly immersive environment which incorporates both physical and digital space. This chapter focuses on the disruptive nature of the metaverse in the workplaces, education, and social lives. It also goes far beyond technology and looks at its human implications, including access, privacy, and equity. With a socio-technical lens, we would emphasise socio-technical co-evolution of capabilities in technology and social practices towards effecting behaviour, learning and social structures in the organisations. The purpose of the chapter is to offer a conceptual background and examine practical change, conduct analysis and assessment of societal, governance, and ethical impacts to the researcher, practitioner, and policymaker [1–4].

14.2 Theoretical and Conceptual Grounding

14.2.1 Metaverse as a Convergence Paradigm: Cyber–Physical–Social Spaces

The metaverse is a convergence of cyberspace, physical reality and social systems into an interactive system of cyberspace called cyber–physical–social spaces. This, unlike analogical digital technologies, enables the sustained creation of virtual worlds that incorporate IoT, digital twins, and immersive areas. Users are able to socialise and transact socially and economically in real-time undermining the online–offline boundaries. This collision has an impact on presence, agency, and identity, making the

T. H. Sardar, *The Spectrum of Computer Science: Emerging Technologies and Trends*, Studies in Smart Technologies, https://doi.org/10.1007/978-3-032-19632-3_14

development of new theoretical frameworks important for gaining an understanding of hybrid experiences [5].

14.2.2 Taxonomy of Core Dimensions: Presence, Immersion, and Interoperability

To systematically review the metaverse experiences, we can consider three fundamental dimensions: presence, immersion, and interoperability [6]. Presence refers to the partial illusion of being in a virtual place, encompassing the conflicts between spatial presence (the sense of being in a virtual place) and social presence (the sense of being associated with other people). It evolves through the interaction of technology and human perception, which can be affected by factors such as visual fidelity, haptics, spatial audio, and avatar embodiment, all of which are strongly correlated with engagement, learning, and emotional attachment. Immersion refers to technical characteristics that can be measured, enabling presence, such as display resolution, field of view, tracking accuracy, and sensory feedback. Whereas presence is considered subjective, immersion gives objective measures, although diminishing gains occur beyond a certain threshold. Interoperability enables platform mobility for users, data, and assets. It encompasses the technical specifications (data formats, communication protocols, identity systems) as well as the governance (terms of service, content moderation, economic exchanges), allowing for a secure, reliable, and interconnected virtual experience in various metaverses.

14.2.3 Frameworks for Evaluating Metaverse Ecosystems

To evaluate metaverse systems, it is necessary to have cross-technical/user-centred/socio-economic assessment frameworks. The technical performance frameworks address scalability, latency, reliability, and security, as these factors determine whether systems are suitable for use at scale. The User Experience Framework analyses the presence, naturalness of interaction, richness of content, and accessibility, considering human-centred design for various users. The Socio-Economic Impact Framework focuses on how the community evolves, economic activity, learning outcomes, and productivity, determining whether the implementation of the metaverse can achieve a substantive value that is not rooted in the novelty of technology [7].

14.3 Enabling Infrastructures and Technologies

14.3.1 XR Interfaces and Interaction Architectures

The principal entry point to the metaverse is Extended Reality (XR) technologies, including VR, AR, and MR, which can determine the availability and practicality of these spaces. VR devices have also improved in terms of resolution, tracking, and wearability, although problems, such as motion sickness and eye fatigue, persistently remain. AR creates difficulties in matching virtual contents with the real world, particularly outdoors, although computer vision and mapping technologies are coming to the rescue. Most importantly, it is intuitive and with hand tracking, voice recognition and natural language helpers, users manipulate objects and interact in a natural way, making XR experiences more life-like and easier to operate [8–10].

As shown in Fig. 14.1, Haptic feedback enhances XR experiences by adding tactile sensations, thereby deepening user immersion. Voice recognition and hand tracking enable natural, hands-free interaction and precise manipulation of virtual objects, respectively. AR and VR interfaces bridge the real and virtual worlds, addressing challenges and advancing technology to shape metaverse environments.

The integration of haptic feedback systems represents a critical frontier for XR interaction design. Current haptic technologies primarily provide limited tactile sensations through vibration, although emerging ultrasound haptics and electrotactile

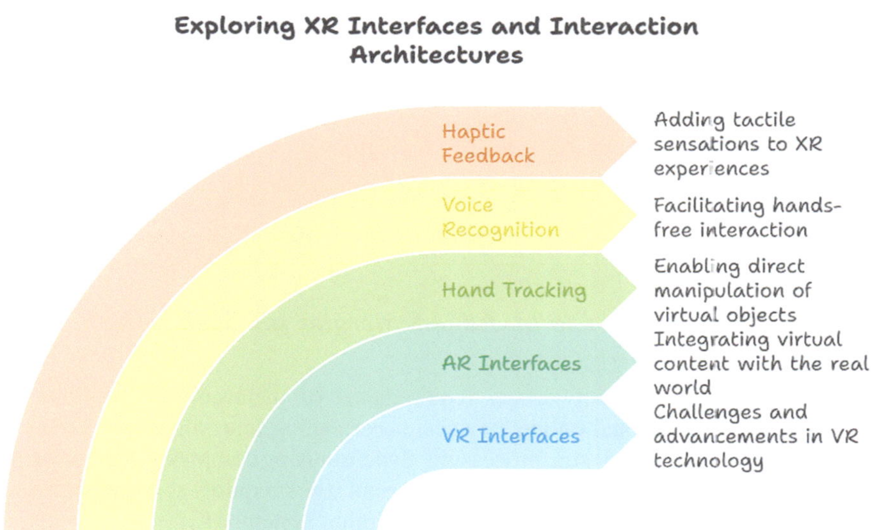

Fig. 14.1 A layered representation of XR interfaces and interaction architectures, illustrating the integration of haptic feedback, voice recognition, hand tracking, AR interfaces, and VR interfaces to enhance user experiences in the metaverse

stimulation show promise for creating more sophisticated touch sensations without physical contact.

14.3.2 Distributed Ledger Technologies and Digital Asset Frameworks

Blockchain and distributed ledger technologies (DLTs) have emerged as foundational infrastructure for metaverse economies, enabling digital asset ownership, decentralised governance, and cross-platform value transfer. The integration of DLTs addresses several critical challenges in virtual world design: establishing verifiable digital ownership, creating interoperable economic systems, and reducing dependence on centralised platform providers [7, 11].

Non-Fungible Tokens (NFTs) represent digital certificates of ownership for unique virtual assets, from avatar clothing to virtual real estate. While NFT markets have experienced significant volatility and criticism regarding environmental impact and speculative behaviour, the underlying technology provides important capabilities for digital asset provenance and transferability.

Smart contracts enable programmable economic relationships within virtual worlds, automating royalty payments to creators, implementing complex governance mechanisms, and facilitating decentralised autonomous organisations (DAOs) that can govern virtual spaces collectively. However, the complexity and immutability of smart contracts introduce new risks related to bugs, security vulnerabilities, and unintended consequences.

The integration of cryptocurrency payment systems enables novel economic models within virtual worlds, including play-to-earn gaming mechanisms and creator economy platforms. These systems face regulatory uncertainty in many jurisdictions and technical challenges related to transaction costs, energy consumption, and scalability.

14.3.3 Network and Cloud–Edge Synergies for Scalability

The computational and networking requirements of metaverse applications far exceed those of traditional online services, necessitating innovative approaches to distributed computing and edge processing. Synchronising complex virtual environments across hundreds or thousands of concurrent users requires sophisticated load balancing, state management, and conflict resolution systems [12].

Cloud computing provides the scalable computational resources necessary for complex physics simulations, AI-driven content generation, and user behaviour analytics that enhance metaverse experiences. However, the latency requirements of immersive applications—particularly for spatial audio and haptic feedback—require edge computing deployment to minimise round-trip times between users and servers.

Edge-cloud hybrid architectures distribute processing between local devices, edge servers, and centralised cloud resources based on computational requirements and latency sensitivity. User interface rendering and basic interaction processing occur locally, while complex simulations and multi-user coordination happen at edge servers, and long-term data storage and analytics occur in centralised cloud facilities.

5G and emerging 6G wireless technologies provide the high-bandwidth, low-latency connectivity necessary for mobile metaverse applications. Network slicing capabilities enable service providers to allocate dedicated network resources for immersive applications, ensuring consistent performance even during peak usage periods.

14.3.4 AI as the Cognitive Layer of the Metaverse

Artificial intelligence serves as the cognitive infrastructure of metaverse systems, enabling intelligent content generation, personalised experiences, and autonomous virtual agents. The integration of AI capabilities transforms static virtual worlds into dynamic, responsive environments that adapt to user behaviour and preferences [13].

Procedural content generation powered by machine learning algorithms can create vast virtual worlds with minimal human artistic input, addressing the scalability challenges of handcrafted virtual environments. Generative adversarial networks (GANs) and diffusion models enable the creation of realistic textures, 3D models, and environmental details that would be prohibitively expensive to create manually.

Natural language processing enables sophisticated conversational interfaces with virtual agents and facilitates real-time translation for cross-cultural collaboration within virtual spaces. Large language models can power virtual assistants, educational tutors, and customer service agents that provide contextually appropriate responses within virtual environments.

Computer vision and machine learning enable automatic content moderation, detecting inappropriate behavior and content within virtual spaces at scale. These systems face significant challenges in understanding context and cultural nuances, particularly when moderating creative expression and social interaction within virtual communities.

Recommendation systems personalise content discovery and social connections within virtual worlds, helping users find relevant experiences and communities. However, these systems raise concerns about filter bubbles and algorithmic bias that could fragment virtual communities and reinforce existing social inequalities.

14.4 Workspaces of the Future

14.4.1 Immersive Collaboration and Productivity Platforms

The transformation of workplace collaboration through metaverse technologies represents one of the most immediate and impactful applications of virtual reality systems. Immersive collaboration platforms transcend the limitations of traditional video conferencing by creating shared virtual spaces where remote team members can interact with digital content and each other in three-dimensional environments [14].

Current immersive collaboration platforms such as Microsoft Mesh, Horizon Workrooms, and Spatial demonstrate significant advantages over traditional video conferencing for certain types of collaborative work. Spatial presence enables natural non-verbal communication through body language and spatial positioning, while shared virtual whiteboards and 3D models facilitate collaborative design and problem-solving activities that are difficult to replicate in traditional digital formats.

The effectiveness of virtual collaboration platforms varies significantly based on task type and team dynamics. Design review sessions, architectural visualisation, and training simulations show particular promise, as these activities benefit from spatial reasoning and shared manipulation of three-dimensional objects. However, text-heavy work and formal presentations may be less suited to current VR interfaces due to resolution limitations and input method constraints.

Research on presence indicates that effective virtual collaboration requires careful attention to avatar design, spatial audio implementation, and shared attention mechanisms. Avatar embodiment significantly impacts social presence and communication effectiveness, with higher-fidelity avatars generally supporting better non-verbal communication but potentially introducing uncanny valley effects that reduce comfort levels.

The integration of artificial intelligence into collaborative platforms enables intelligent meeting assistance, automatic note-taking, and real-time language translation that can enhance productivity and accessibility. AI-powered virtual assistants can manage scheduling, resource allocation, and task tracking within virtual workspaces, reducing administrative overhead and enabling focus on creative and strategic activities.

14.4.2 Enterprise Digital Twins for Simulation and Decision-Making

Digital twin technologies represent a convergence of IoT sensors, real-time data processing, and virtual reality visualisation that enables organisations to create virtual replicas of physical assets, processes, and systems. Within metaverse contexts, digital twins become interactive, collaborative environments where stakeholders can explore complex systems, test scenarios, and make data-driven decisions [15].

Manufacturing organisations have pioneered digital twin applications for optimising production lines, predictive maintenance, and quality control. Virtual representations of manufacturing facilities enable remote monitoring, collaborative troubleshooting, and training simulations that reduce downtime and improve safety outcomes. Companies like Siemens and General Electric report significant cost savings and efficiency improvements through the implementation of digital twins.

Urban planning and smart city initiatives increasingly rely on digital twin platforms that integrate transportation data, environmental sensors, and population dynamics into comprehensive virtual city models. Digital twin platforms can be useful in simulating real-world systems, interacting with the stakeholders, and improving decisions. They are used in urban planning to visualise suggested changes, engage citizens and to streamline resource utilisation across significant infrastructures. Patient specific models are used in healthcare to aid surgery planning; medication testing; or in simulation of healthcare; in teaching patients through the simulations. To decide correctly on these virtual tools, success depends on real time data that is accurate, handy interfaces, and good training and ensure that the virtual tools make these decisions easier instead of becoming more complicated.

14.4.3 Human–AI Synergy and Adaptive Workflow Automation

The concept of the metaversion workspace enabled through AI integration enables human and AI to work together to produce enhanced outcomes based on how human creativity interacts with AI quantitative processing capabilities. The adaptive workflow automation is capable of monitoring the behaviour of the user, detecting patterns and proposing even the improvement without taking away human control. Smart virtual assistants will offer contextual aids, schedule management, sharing of knowledge, which learns through interaction thus becoming more personalised without infringing privacy [16].

AI-assisted tools like prototyping and ideation can also be used to implement numerous solutions in a short amount of time. Predictive analytics can foresee problems, future resource utilisation and provide process recommendations, particularly under difficult environments where humans find it difficult to observe all variables.

Nevertheless, the use of AI is associated with concerns such as job displacement, bias in the algorithm, and a loss of autonomy. Trust is only maintained when transparency, explainability, and control by the users are upheld. Human-AI synergy, when applied meaningfully, can also help simplify operations, improve decision makers and creativity, without dislodging human judgement.

Figure 14.2 illustrates that the IoT sensor integration and real-time synchronisation provides a seamless flow of data into the digital twin, which is the cornerstone thereof.

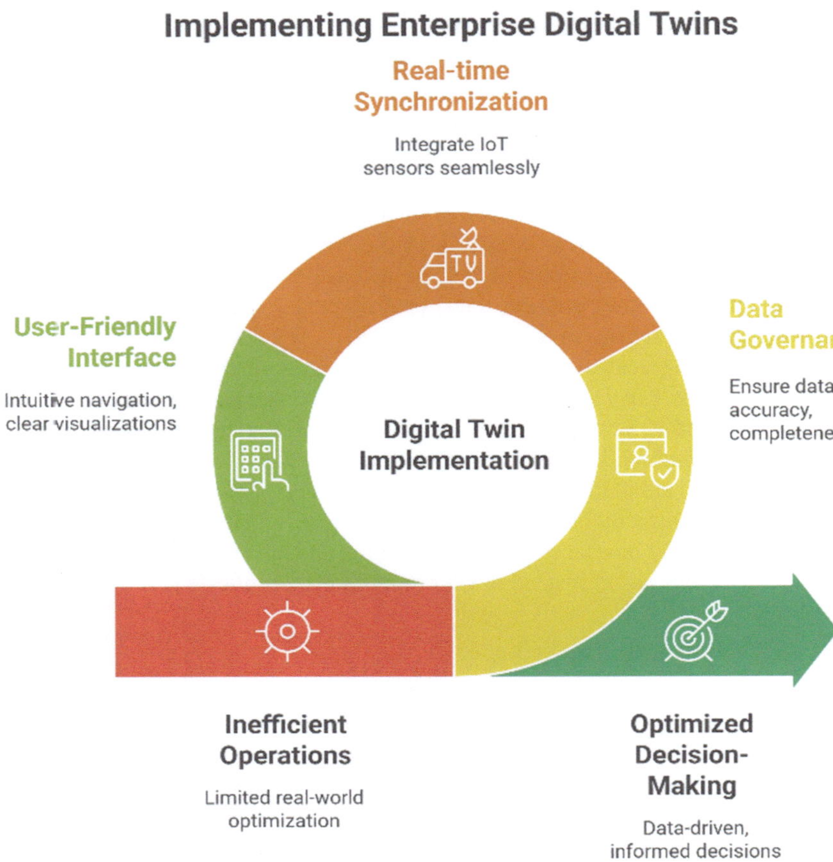

Fig. 14.2 A cyclical model of implementing enterprise digital twins

14.4.4 Organisational Adoption Models and Readiness Indicators

The use of metaverse technologies is conditioned by technical preparedness of an organisation, organisational culture, and the external forces. Technically advanced firms (which operate internationally and can have distributed teams) tend to be the first ones to use new collaboration tools [17].

IT infrastructure, digital literacy of employees, leadership support, and being open to change all shape organisational readiness. The initial process of successful adoption is to set up pilot projects and then perform a gradual extension, which is measured by feedback information. Introduction of training, technical assistance and change management are important in ensuring that resistance is addressed and the use is effective.

The cost–benefit analysis should take into account direct expenses (equipment, software), as well as indirect ones (training, downtime on work, etc.), and weigh them in the balance against the gains in terms of decreased travel, increased cooperation, and exposure improvement. The difference in ROI depends on the size of an organisation, its complexity and quality of implementation. Careful planning helps to make immersive technologies provide substantial value instead of being an innovative technological trend.

14.5 Reimagining Education in the Metaverse

14.5.1 Metaversal Learning Ecosystems: Virtual Classrooms to Knowledge Metaspaces

Metaverse learning brings education outside the virtual classrooms in fascinating and participative ecosystems. Students have the ability to manipulate with 3 dimensional models, do work together or work on projects and also get involved into an immersion simulation—be it in exploring the old civilisation, studying a molecular structure or navigating a solar system amongst others. They result in hands-on experiential learning, which cannot be accomplished in the traditional environments [18].

The metaspaces of knowledge allow a deep exploration of subject areas in the form of interactive virtual worlds, discovery-based learning, collaborative learning, and long projects. Individualisation of learning paths is supported by AI integration, gaps in learning are detected, and feedback is generated, instructors also target a teaching-focused approach and social–emotional support.

In these environments, assessment is no longer differentiated by tests but rather authentic performance-based assessment of complex skills and ability. Combining VR-based experiences with AI-guided mechanisms means that metaversal ecosystems allow students to engage in meaningful personalised learning on a large scale, and the collaboration and social aspects involved are equally important.

Students can demonstrate understanding through the creation of virtual artefacts, the successful completion of simulated scenarios, and collaborative problem-solving activities that provide rich behavioural data for learning analytics systems.

14.5.2 Gamification, Experiential Learning, and Neuroadaptive Pedagogy

The immersive nature of metaverse environments naturally supports gamification strategies that increase student engagement and motivation through game-like mechanics, narrative structures, and achievement systems. However, effective educational gamification requires careful balance between entertainment value and learning objectives to avoid superficial engagement that doesn't translate to meaningful learning outcomes [19].

Quest-based learning structures enable students to pursue personalised learning goals through sequences of interconnected challenges that build skills progressively. These systems can incorporate branching narratives that adapt to student choices and performance, creating individualised learning experiences within shared virtual environments. Collaborative quests encourage teamwork and peer learning while individual challenges allow for self-paced skill development.

Experiential learning within virtual environments enables students to engage with dangerous, expensive, or impossible scenarios safely and repeatedly. The metaverse technologies are transforming the educational system by allowing students to learn via 3D experience, which is risk-free. Med students are able to perform surgeries, engineers are able to survive inharsh conditions when testing their designs, and students in social sciences are able to learn in situ. Repeat and reset: Mastery learning and skill development is supported by the ability to repeat and reset experiences which is not achievable in traditional classrooms.

Few personalisation approaches are extended to neural adaptive their pedagogy which uses technologies such as eye tracking or EEG to monitor lapses or confusion of attention and modify content or pace accordingly. As radical as they are, these ways need the consideration of privacy and morality.

Through the PESTEL complexity illustrated in Fig. 14.3, fundamental challenges have been identified: political and accreditation concerns, digital equity, cultural inclusivity, adaptive technologies, energy sustainability, and Internet privacy. Careful consideration of these factors is able to render immersive education as mind-engaging and valuable as can be serious concerns to learners across the globe.

14.5.3 Accessibility, Inclusivity, and Cognitive Diversity in Learning

There are great opportunities to use metaverse based learning environments to realise education accessibility and inclusivity, yet associated challenges are specific. Virtual classrooms have the ability to be modified to suit the varying abilities and learning styles in a manner that physical space does not always. In high-contrast visuals, adjustable text, colour differentiation, and alternative content formats, designers can modify these features so that they can be applied to visually impaired learners. Spatial audio, haptic feedback, voice recognition, and text-to-speech also reduce accessibility, but existing VR equipment does not always offer the option of all features [20].

For users with restricted mobility, virtual learning can reduce the physical limitations by providing tailor-made controls and eliminating movement demands, allowing users to experience what they would not otherwise be capable of in the real world. Cognitive accessibility is also of critical relevance: a carefully designed

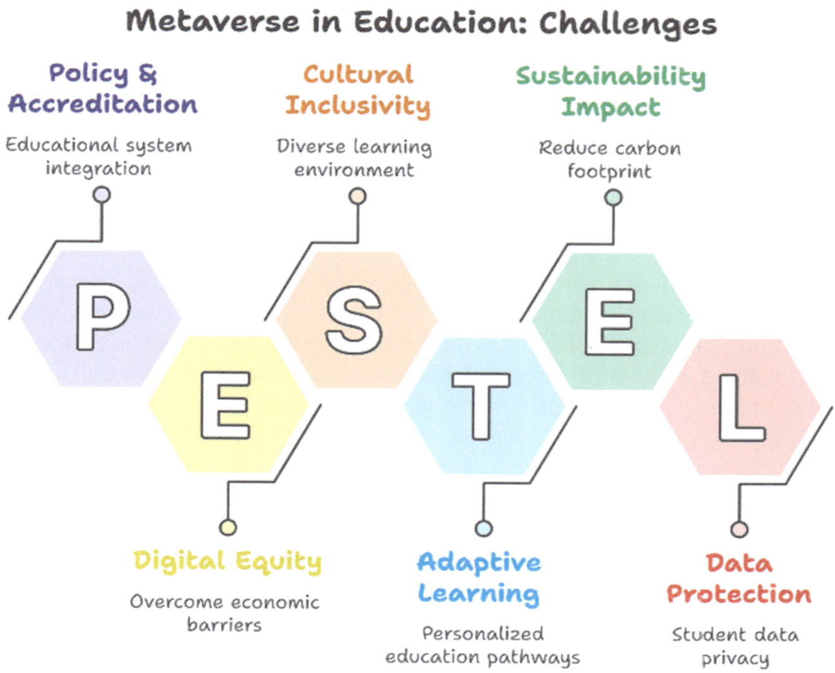

Fig. 14.3 A PESTEL framework analysing challenges of the metaverse in different sectors

interface, quality navigation, attention, task, and note-taking facilitation support tools can be used to empower learners with processing variations to succeed.

Cultural inclusivity that takes into account the different norms, communication styles, and educational traditions should be practiced. The images and content should not be generic or insensitive, and they should reflect the diversity of the world. Economic accessibility is also a problem because high-quality immersive learning occasionally necessitates costly equipment and fast Internet. It is essential to bridge this digital divide so that metaverse learning can mitigate existing educational inequities rather than exacerbating them.

14.5.4 Challenges in Accreditation, Ethics, and Digital Divide

In their technology acquisition, finding a proper balance between educational innovation and environmental responsibility means paying close attention to energy conservation and carbon footprint considerations [21].

14.6 Socio-cultural Dynamics and Human Experience

The adoption of metaverse technologies in the formal education process raises critical questions regarding accreditation, ethics, and equity. Critical assessment procedures tend to fail when immersive learning is employed; therefore, ascertaining new quality criteria through established standards is essential to deliver quality and validity. One of the problems is proving that skills learnt online are also applicable to the real world, since they have different degrees of effectiveness across subjects and situations [22].

Another major issue is student privacy. By their nature, metaverse stages gather more behavioural data, which increases the risks involved with surveillance, manipulation, and consent. Schools need to implement open data management policies and privacy consent solutions to safeguard the freedom of students. There are also ethical issues surrounding the influence of virtual environments that may solely persuade behaviours even without intentional agreement.

Digital equity is not merely a matter of access to hardware, but also takes into account the cost of technical support, digital literacy, and the cultural experience of immersion in technology. Some students may have cultural or religious concerns about virtual reality, and others who have had little exposure to it may need extra help. In conclusion, careful consideration must be given to the environmental sustainability of power-consuming systems, such as blockchain integration or high-performance computing. To guarantee that all students gain something from metaverse education, it is necessary to strike a balance between innovation, equity, and sustainability as the metaverse expands.

14.6.1 Communication Paradigms and Collective Presence

Through the introduction of new interaction modalities that creatively combine verbal, non-verbal, and environmental cues, the metaverse radically transforms human communication. Metaverse environments facilitate embodied communication, in contrast to conventional digital communication platforms that mainly use text or video. This allows users to express themselves through the movement of their avatars, the placement of their avatars in space, and the manipulation of virtual objects.

When human relationships and technological features interact, community presence—the shared sense of "being together" in a virtual space—is developed. Research indicates that collective presence is enhanced by synchronised activities, shared attention toward common objects, and the ability to maintain eye contact and spatial relationships similar to physical interactions. However, the mediated nature of virtual interaction introduces new challenges for establishing trust, reading social cues, and maintaining authentic relationships.

Virtual embodiment through avatars creates complex psychological dynamics where users may experience heightened self-awareness, altered self-perception, and modified social behaviour. The "Proteus effect" demonstrates that avatar appearance and characteristics can influence user behaviour and self-concept, with implications for identity formation and social interaction within virtual environments.

Cross-cultural communication within global metaverse platforms requires navigation of diverse communication styles, cultural norms, and linguistic differences. Real-time translation technologies enable basic communication across language barriers, but cultural nuances in non-verbal communication, humour, and social etiquette require more sophisticated cross-cultural competency development.

The persistence of virtual spaces enables the development of shared histories, cultural artefacts, and community traditions that contribute to social cohesion and collective identity formation. Virtual communities can develop their own norms, rituals, and social structures that may differ significantly from physical world conventions.

According to Fig. 14.4, the metaverse provides virtual embodiment, helping shift self-perception and behaviour, whereas collective presence helps to achieve a sense of togetherness. Embodied communication provides powerful and in-depth experiences whereas the cross-cultural communication evaluates different cultural norms. Online communities establish their own social norms and networks, and establish new ways of interaction, creating innovative social systems.

14.6.2 Digital Identity, Representation, and Authenticity

Because they can experiment with gender, age, appearance, and even character—things they can never be in the real world—users of avatars are able to liberate themselves. When people start to wonder more about their identities or how to overcome physical biases, this overdevelopment can be freeing.

Users' preferences vary; some choose fantastical or idealised avatars, while others use ones that closely resemble their real-life appearance. These choices have an impact on how other people view and treat them, which somewhat reflects the prejudices expressed online. By implementing identity or reputation verification systems, platforms can achieve a balance between privacy, self-expression, and trust.

Avatars, reputations, and social connections are all investments in virtual identities. It is essential that these identities persist, be portable, and find respect across digital spaces as users live, interact, and make meaning on the internet in more and more ways.

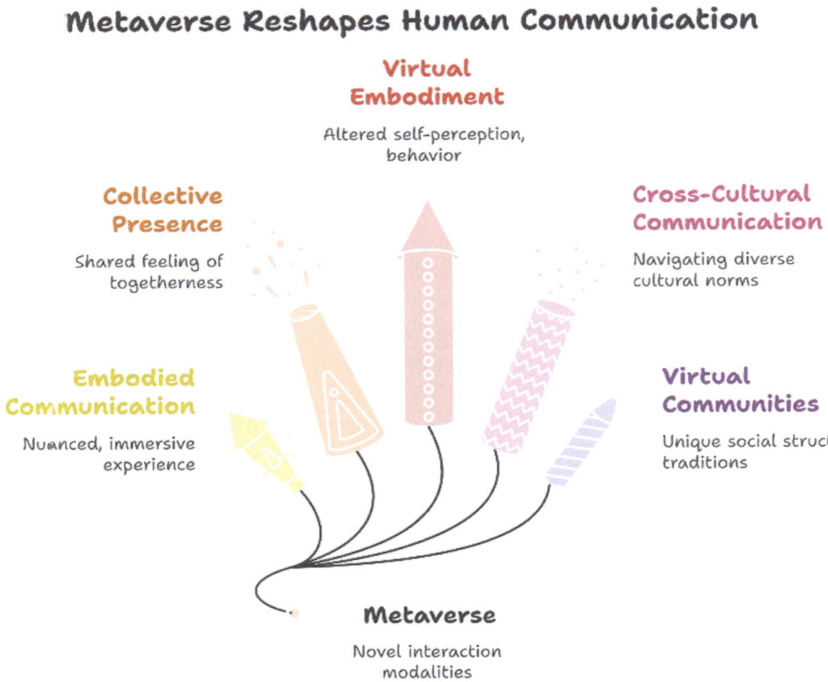

Fig. 14.4 The metaverse reshapes human communication through virtual embodiment, shared presence, cross-cultural interaction, and immersive communities

14.6.3 Cultural Exchange, Virtual Economies, and Community Resilience

Cross-cultural exchange can be seen to increase through metaverse platforms, where individuals from diverse backgrounds can engage with one another, learn new languages, and create global communities. Healthy interactions, though not without some possibility of misunderstanding, can result in understanding or conflict.

Using virtual economies, users can create, trade, and monetise digital properties, including virtual real estate and fashion accessory materials, events and content creation, which provide alternative sources of income, particularly because not everyone has access to comparable opportunities in the real world. Democratisation of the creation tools allow that more individuals can contribute without the need to have technical skill or vast investments which enhances the cultural representation and creativity.

According to community resilience, it is based on quality governance, conflict resolution and values. Serious virtual communities tend to have a proper leadership, moderating, and making decisions together in order to have friendly circumstances.

Nevertheless, there is a problem of the predominance of such platforms worldwide, making it difficult to conduct without taking into account cross-jurisdictional legislation and to avoid the imposition of various cultural standards.

14.6.4 Risks of Alienation, Well-Being, and Cyber-Social Harms

There is unreliable information on the subject of trespass by third parties as some maternal cases of trespass are part and parcel of Californian communities.

Spending time in the immersion of virtual worlds can have mental health consequences, social relationships, and obligations. Although virtual interactions can supplement physical world relationships, excessive use is likely to lead to physical world avoidance, neglect, and isolation.

These environments enable immersion and thus seem more gratifying than reality, which can further complicate social and psychological problems, particularly for individuals who feel safer behind the computer screens. Attacks and harassment via cyberbullying have the potential of being more personal and coordinated with visual and spatial immersion.

Addictive behaviours may be driven by gamification, social pressure and the desire to miss out, which raises an ethical concern about the responsibility of the platforms and the necessity to monitor use.

Another issue is privacy, which could be abused to manipulate or discriminate, as detailed behavioural tracking, including eye movements, gestures, and emotions, could be used.

Last but not least, engaging with different web identities is beneficial in the process of self-formation, even though it may be deceitful or trigger self-inflation, particularly for younger users who have not fully come to terms with who they are.

14.7 Opportunities, Risks, and Governance Landscapes

A thorough investigation of value development, transfer, and authority is necessary due to the profound economic changes brought about by the metaverse in addition to the changes in social interaction [23].

14.7.1 Economic Value Chains and Innovation Pathways

The potential of investors, regulators, and creative individuals is being bound by the new value chains that the metaverse markets are creating in hardware, software, content, digital services, and other electronic assets.

Supply chain risks arise from the hardware, such as VR/AR devices, haptic systems, and high-performance computing, which is often concentrated in small spaces. The ecosystem is dominated by software platforms, but interoperability standards have the potential to encourage healthier competition.

Using next-generation content creation, individuals and small groups can monetise the virtual experiences and services with the help of safe revenue sharing, effective tools and favourable platform policies. Virtual services, ranging from education to entertainment, can reach a global audience, but these are raising questions about the protection of labour.

Decentralised, blockchain models allow creators to connect with consumers directly; however, cryptocurrency uncertainty and complexity create barriers to adoption, as well as economic fees.

14.7.2 Regulatory Regimes, Standards, and Global Governance

Our current judicial system lacks the ability to confront the interconnected immersiveness of the metaverse, leading to a regulatory obstacle. By crossing limits, computer networks reveal consequences that the systems of law which they're comprised of could fail to account for, as well as creating new identities and properties.

Text filters are just one part of content moderation; other problems include cultural diversity, tolerance for free speech, and visual, spatial, and behavioural issues. For the automated systems to distinguish between acts of violence and genuine creativity, context sensitivity is required.

Privacy laws should also be updated because immersive platforms record biometric and behavioural data that is far beyond the scope of laws like the GDPR. The intellectual property systems must also resonate with this, which means that the ownership and protection of 3D objects, user-generated content, and collaborative works should be treated fairly.

Virtual worlds do not respect jurisdictions, and therefore, international collaboration is needed. To achieve an open, safe, and fair metaverse, governments, technology companies, and civil society organisations must work together to establish a set of global governance, interoperability, and security rules.

14.7.3 Responsible Innovation and Ethics-by-Design Frameworks

Ethics are a crucial focus in the construction of the metaverse so that it can contribute to positive human lives instead of strengthening injustices or facilitating exploitation. Development must be designed with ethics as it should be more fair and responsible at every phase.

The artificial intelligence used to facilitate suggestions, forecasts, and editing should be informative, and provide visuals, as it will unconsciously circulate prejudice or influence user behaviours destructively. Consent processes, on the same level, should be modified, as the traditional method of clicking through is insufficient to describe the way in which immersive platforms gather and process biometric, behavioural, and social data.

The users also have a right to privacy in the digital format that acknowledges emotional and monetary value of their virtual property, avatar identity, and involvement in online communities. Finally, the active involvement of stakeholders, namely, various communities, professionals, and civil societies, is essential to make the metaverse develop in a manner that would represent the general interests of the society rather than those of companies only.

14.7.4 Sustainability, Environmental Impact, and Energy Considerations

Sustainable development is highly concerned with the implications of the metaverse to the environment since it consumes a lot of energy to run both immersive virtual worlds, compared to just running the digital service. The use of high-performance graphics, real time rendering as well as blockchain increases energy requirements and carbon emissions. With increased adoption, data centres will also demand excessive energy, and improved energy-efficient performance and integration of renewable energy should be critical. The environment imprint also hinges on the placement of these centres. Those connected with fossil-fuel grids have a greater impact on the environment preciosity compared to those connected with clean energy. Together with operational issues, device production contributes to the difficulty as it requires rare earth mining, consumes high energy at the manufacturing level, and exacerbates the growing problem of e-waste due to ongoing hardware upgrades. Commodities that are more durable, repairable, and recyclable will play a crucial role in maintaining this weight.

Carbon offset and sustainability initiatives within metaverse platforms can include virtual environmental education, digital-first alternatives to physical activities, and gamification of sustainable behaviour in physical contexts. Some platforms are exploring "green virtual worlds" that promote environmental awareness and sustainable practices among users.

Life cycle assessment methodologies must be adapted to evaluate the comprehensive environmental impacts of metaverse systems, including direct energy consumption, indirect effects from reduced travel and physical space requirements, and potential rebound effects from increased overall technology usage.

14.8 Foresight: Metaverse Futures

Looking beyond present applications, the metaverse points toward transformative futures where digital and physical realities increasingly converge [24, 25].

14.8.1 Evolution Towards Mixed-Reality Societies

The trajectory of metaverse development points toward the emergence of mixed-reality societies where the boundaries between physical and virtual experiences become increasingly blurred and irrelevant to daily life. This evolution represents a fundamental shift in how humans relate to space, objects, and each other in an augmented world where digital information and virtual objects are seamlessly integrated with physical environments.

Augmented reality contact lenses and retinal projection systems may eventually eliminate the need for separate devices to access virtual content, enabling continuous augmentation of physical environments with contextual information, social connections, and digital services. This ubiquitous computing vision would transform every physical space into a potential interface for virtual interaction and information access.

With the convergence of the Internet of Things (IoT) with the metaverse, commonplace objects, such as homes, cars, city infrastructure, and so forth, might acquire virtual twins that we can interact with and personalise using immersive interfaces. Such convergence between the physical and the digital creates possibilities for completely new methods of communication with the environment. Brain–computer interfaces offer the idea that human consciousness can directly interact with the virtual world. Brain–computer interfaces are in the early stages of research, with the potential for seamless integration between human consciousness and the brain, which could be directly utilised for entering the virtual world. The mixed reality community divides human beings into augmented and non-augmented individuals, providing a quality of privacy and solitude, while reducing our physical interaction with the world as we traditionally do. Researchers expect that this may create new words of creativity that we cannot envision now.

14.8.2 Metaverse for Sustainable Development and Global Cooperation

The metaverse has enabled humankind to come together while erasing the divide in terms of border lines, citizenship, and pressing common issues. Metaverse-based virtual environments are providing pathways that traditional institutions cannot provide. For example, the concept of Climate Change and its impact on the global environment can be better understood through immersive visualisations. This visualisation can provide an easy understanding of the global ecosystem, its changing nature due to climate change and underlying factors, and its impact on human lives. You can further help us understand the need for renewable energy and other environmentally friendly techniques. Similarly, virtual simulations can help professionals working in the Healthcare domain to get trained without real-world access to medical equipment. Immersive education provides students with the opportunity to learn about environmental sustainability, disease prevention and treatment, prisons, and many other useful concepts across various domains.

Telemedicine, which is offered in virtual platforms, has the potential to expand healthcare to disadvantaged areas. Education, too, will be a winner: virtual schools and universities will be able to provide quality education to students regardless of their geographical location, but it will be essential to provide digital access and supplement local institutions. To scientists, virtual laboratories and research facilities enable them to collaborate with others around the world to solve complex issues, and virtual conferences eliminate the need to travel or produce emissions, allowing many who were previously shut out to participate.

14.8.3 Long-Term Socio-economic Transformations and Open Research Questions

The proliferation of the metaverse will not only change technology but the society as a whole. The work world can be more global since immersive platforms destroy the physical geographical boundaries, but this can undermine local economies and social connections. Education might become less classroom oriented and more interactive and experiential education, and it may question the old system of certification and authority.

Governments are currently grappling with the issue of policing virtual spaces that are available to individuals in various countries and continents, which play a part in the economy of the government and create virtual communities that offer users a sense of being citizens of that community. Metaverse-based interdisciplinary research projects must be invested by the government and enterprises. To illustrate, some of the issues that researchers can address include the effects of virtual life on individual mental and physical health, its influence on cultural and societal norms, sustainability, and the authenticity of virtual economics, among others. The issues also raise a vital question concerning the disparity between the online users: Can it be extended to assist diminish it?

14.9 Conclusion

The metaverse is not a password within the computer science world. It is changing the way we are handling modern infrastructure, which is what we study in the technology field and interact with other human beings. In this chapter, a review of the issues has been given with emphasis on the challenges that it is experiencing.

Some of these themes are significant within a given domain and context. To illustrate, virtual experiences can be used in a meaningful way when they are supplemented by actual human relationships and their presence. The metaverse technology is heavily contributed by artificial intelligence. It must not eliminate human creativity, but increase it. Then there are the cultural and social factors to be considered as important as the technical factors of the metaverse. Our platform should be the one that makes us feel like we are part of one community. Finally, the metaverse must be created in a manner that is accommodative to every field and type of individuals.

In the future, researchers ought to explore the potential use of immersive technology in the long run to positively influence well-being, learning, and community life in different cultures and environments. The practitioners are expected to design with the users providing the training and support in mind, whereas policymakers must struggle with the difficult task of ensuring privacy, equity, and competition without stifling innovation. The key to a sustainable and equitable metaverse will be open standards and investment in digital accessibility.

The metaverse is not doomed to the future. It will be influenced by our decisions today, whether to utilise it to support creativity, inclusion and social progress, or whether to allow it to become an additional commercial playground. When we think carefully and creatively about it and do so collectively, the metaverse can become a place where human potential truly grows.

References

1. Das SK. Cyber-physical-social convergence in smart living: challenges and opportunities. In: 2016 IEEE international conference on pervasive computing and communication workshops (PerCom Workshops). Sydney, NSW, Australia; 2016, p. 1–1. https://doi.org/10.1109/PERCOMW.2016.7457093.
2. Ding K, Jiang P. Incorporating social sensors, cyber-physical system nodes, and smart products for personalized production in a social manufacturing environment. Proc Inst Mech Eng Part B: J Eng Manuf. 2017;232(13):2323–38. https://doi.org/10.1177/0954405417716728.
3. Xu X, Chen D, Cheng TC, Choi TM. When should metaverse technology be adopted in live streaming considering consumer returns? Eur J Oper Res. 2025. ISSN 0377-2217. https://doi.org/10.1016/j.ejor.2025.08.019.

4. Singh R, Sarkar P, Bandyopadhyay A, Swain S, Sardar TH, Pandey B. Combinatorial kinetic dualistic auction model for metaverse service allocation. In: 2024 IEEE AITU: digital generation; 2024. IEEE. p. 27–32.
5. Shi F, Ning H, Zhang X, Li R, Tian Q, Zhang S, Zheng Y, Guo Y, Daneshmand M. A new technology perspective of the metaverse: its essence, framework and challenges. Dig Commun Netw. 2024; 10(6):1653–1665. ISSN 2352-8648. https://doi.org/10.1016/j.dcan.2023.02.017.
6. Sadeghi-Niaraki A, Rahimi F, Azlan NAEB, et al. Groundbreaking taxonomy of metaverse characteristics. Artif Intell Rev. 2025;58:244. https://doi.org/10.1007/s10462-025-11243-5.
7. Tsappi E, Deliyannis I, Papageorgiou GN. Developing a performance evaluation framework structural model for educational metaverse. Technologies. 2024; 12(4):53. https://doi.org/10.3390/technologies12040053
8. Mula J, Sanchis R, de la Torre R, Becerra P. Extended reality and metaverse technologies for industrial training, safety and social interaction. IFAC-PapersOnLine. 2024; 58(19):575–580. ISSN 2405-8963. https://doi.org/10.1016/j.ifacol.2024.09.274.
9. Guan J, Liu J, Morris A. Design frameworks for spatial zone agents in XRI metaverse smart environments. In: 2024 IEEE international conference on artificial intelligence and eXtended and virtual reality (AIxVR). Los Angeles, CA, USA; 2024. p. 75–79. https://doi.org/10.1109/AIxVR59861.2024.00017.
10. Sonkamble RG, Shirke-Deshmukh S, Katkar V, et al. A blockchain secured metaverse framework for scalable and immersive telemedicine. Sci Rep. 2025;15:26655. https://doi.org/10.1038/s41598-025-12068-6.
11. Ahmad A, Shah A, Alnumay WS, Adnan M, Anwer S, Zaman QU. A comprehensive survey on the convergence of blockchain, digital twins, and metaverse: shaping the future of cybersecurity frameworks. Comput Electr Eng. 2025; 126, 110486. ISSN 0045-7906. https://doi.org/10.1016/j.compeleceng.2025.110486.
12. Moro-Visconti R, Cesaretti A. The metaverse. In: Digital token valuation. Cham: Palgrave Macmillan; 2023. https://doi.org/10.1007/978-3-031-42971-2_7
13. He B, Alshaghdali NO, Ren Y, Romano M, Cuomo MT. Augmenting user experience in the AI-transitioned metaverse: the role of knowledge management in socio-economic inclusion. Technol Soc. 2025; 83, 102996. ISSN 0160-791X. https://doi.org/10.1016/j.techsoc.2025.102996.
14. Dwivedi et al. Metaverse beyond the hype: Multidisciplinary perspectives on emerging challenges, opportunities, and agenda for research, practice and policy. Int J Inf Manag. 2022; 66, 102542. ISSN 0268-4012. https://doi.org/10.1016/j.ijinfomgt.2022.102542.
15. Qu M, Yao Y, Sun M, Chang X, Li Q. Industrial digital twins based on enterprise modeling: architecture, methodology, and engineering applications. J Ind Inf Integr. 2025; 47, 100912. ISSN 2452-414X. https://doi.org/10.1016/j.jii.2025.100912.
16. Cao L. Ai4tech: X-AI enabling X-Tech with human-like, generative, decentralized, humanoid and metaverse AI. Int J Data Sci Anal. 2024;18:219–38. https://doi.org/10.1007/s41060-024-00615-9.
17. Singh A, Luthra A, Garg S, et al. Banking transformation in PSU banks through the adoption of metaverse: Indian context. Int J Syst Assur Eng Manag. 2025. https://doi.org/10.1007/s13198-024-02611-5.
18. Gurkan AM. Enhancing education through the metaverse: perspectives from Türkiye. Digit Soc. 2025;4:31. https://doi.org/10.1007/s44206-025-00183-w.
19. Sandrone S, Carlson C. Gamification and game-based education in neurology and neuroscience: applications, challenges, and opportunities. Brain Disorders. 2021; 1, 100008. ISSN 2666-4593. https://doi.org/10.1016/j.dscb.2021.100008.
20. Zallio M, Clarkson PJ. Designing the metaverse: a study on inclusion, diversity, equity, accessibility and safety for digital immersive environments. Telem Inform. 2022; 75, 101909. ISSN 0736-5853. https://doi.org/10.1016/j.tele.2022.101909.
21. Tabassum A, Ghaznavi I, Abd-Alrazaq A, Qadir J. Exploring the application of AI and extended reality technologies in metaverse-driven mental health solutions: scoping review. J Med Internet Res. 2025; 27. ISSN 1438-8871. https://doi.org/10.2196/72400.

22. Damodaram D, Sundar S, Kumar NA. Unveiling the metaverse: a paradigm shift in the future of internet interaction. In: 2023 IEEE fifth international conference on advances in electronics, computers and communications (ICAECC). Bengaluru, India; 2023, p. 1–6, https://doi.org/10.1109/ICAECC59324.2023.10560205.
23. O. da Silva MM, XN Teixeira JM, FF Peres F et al. Learning in the metaverse: reflections on potential benefits, possibilities and challenges. Digit Soc. 2025; 4, 60. https://doi.org/10.1007/s44206-025-00217-3.
24. Ristola J. Queering the metaverse: envisioning the future of embodiment in VR. Digit Soc. 2025;4:43. https://doi.org/10.1007/s44206-025-00199-2.
25. Pierucci F. "The Land on Which We Meet"—Modes of indigenous self-determination in the metaverse. Digit Soc. 2025;4:36. https://doi.org/10.1007/s44206-025-00182-x.

MIX
Papier aus verantwortungsvollen Quellen
Paper from responsible sources
FSC® C105338

FSC
www.fsc.org

If you have any concerns about our products,
you can contact us on
ProductSafety@springernature.com

In case Publisher is established outside the EU,
the EU authorized representative is:
Springer Nature Customer Service Center GmbH
Europaplatz 3, 69115 Heidelberg, Germany

Printed by Libri Plureos GmbH
in Hamburg, Germany